The utilization of unconventional microbial sources, particularly microalgae, for the production of feed, food, food additives, pharmaceuticals, and fine chemicals is growing in importance. Research in the field is expanding worldwide.

The author presents a state-of-the-art account of research in algal production and utilization. The book explores in detail all steps of the subject, from the preparation stock cultures to the growth in large outdoor ponds. Dr Becker provides a compilation of the different methods employed worldwide for the artificial cultivation of different microalgae, including recipes for culture media, descriptions of outdoor and indoor cultivation systems, as well as harvesting and processing methods. The book will be essential reading for advanced undergraduates, postgraduates and researchers in the field.

T0245150

Cambridge Studies in Biotechnology

Editors: Sir James Baddiley, N.H. Carey, I.J. Higgins,
W.G. Potter

10   Microalgae: biotechnology and microbiology

10 Microalgae: biotechnology and microbiology

Other titles in this series

Microalgae

Biotechnology and microbiology

E.W. BECKER
Medical Clinic, University of Tübingen

CAMBRIDGE
UNIVERSITY PRESS

CAMBRIDGE UNIVERSITY PRESS
Cambridge, New York, Melbourne, Madrid, Cape Town, Singapore, São Paulo

Cambridge University Press
The Edinburgh Building, Cambridge CB2 8RU, UK

Published in the United States of America by Cambridge University Press, New York

www.cambridge.org
Information on this title: www.cambridge.org/9780521350204

First published 1994
Reprinted 1995
This digitally printed version 2008

A catalogue record for this publication is available from the British Library

Library of Congress Cataloguing in Publication data

Becker, E.W. (E. Wolfgang), 1939–
Microalgae: biotechnology and microbiology / E.W. Becker.
    p.   cm. – (Cambridge studies in biotechnology)
Includes bibliographical references and index.
ISBN 0-521-35020-4
1. Algae – Biotechnology. I. Title.   II. Series.
TP248.27.A46B43   1993
660′.62–dc20   93-2129   CIP

ISBN 978-0-521-35020-4 hardback
ISBN 978-0-521-06113-1 paperback

Contents

1 Introduction

The large-scale cultivation of microalgae and the practical use of its biomass as a source of certain constituents (for instance, lipids) was probably first considered seriously in Germany during World War II. This initial research was taken up by a group of scientists at the Carnegie Institution of Washington, who summarized their experiences in the classic report *Algal culture from laboratory to pilot plant* (Burlew, 1953). The aim of this project was the use of the green alga *Chlorella* for large-scale production of food, based on the uncertainty of whether the yields obtained under laboratory conditions could be maintained outdoors on a larger scale.

In the course of time, the continuous cultivation of algae under partially or fully controlled conditions has become an important development, with various economic possibilities. The past decades have witnessed great strides in the production and utilization of algae through the efforts of many scientists in several countries. Techniques for the cultivation of algae on a large scale and proccesses for their utilization have been successfully developed in several countries and an attempt in this direction is worthwhile in some of the developing countries (Venkataraman & Becker, 1985).

During this early period of algal mass cultivation, technological advances and nutritional evaluations have received major attention. However, widespread criticism on the exploitation of unconventional protein sources in general, and microbial proteins in particular, has affected the aspirations relating to the use of algae as supplementary food for human consumption. The great expectations promised initially in the use of algae as food have subsided for several reasons, and today researchers seem to be more cautious in expressing any opinion about the practicability of algal production on a commercial scale for use in foods.

This change in attitude is mainly a result of the high cost of algal production, acceptability problems, and safety criteria. Consequently, extensive research on different fundamental and applied aspects of phycology were carried out that demonstrated that algal biomass can be used for various other applications such as animal feed, biofertilizer, soil conditioner, and as feed in aquaculture, and that it may help to solve public health problems by means of biological purification of waste waters of a fast-developing society. Latest developments have established the potentiality of algae for the production of a variety of compounds such as polysaccharides, lipids, proteins, carotenoids, pigments, vitamins, sterols,

1

Table 1.1. *Projected applications of commercially produced algae*

I. Food	Protein supplement/fortification in diets for malnourished children and adults.
II. Feed	Protein/vitamin supplement in feeds for poultry, cattle, pigs, fish and bivalves.
III. Health food	Algal powder as ingredient and supplement in health food recipes and products.
IV. Therapeutics	β-Carotene as possible anti-skin-cancer treatment. Algal antibiotics as wound treatment, enzymatic hydrolyzates to promote skin metabolism. Prostaglandin stimulation by γ-linolenic acid. Regulation of cholesterol synthesis. Isotopic compounds in medical research.
V. Pigments	β-Carotene as food color and food supplement (provitamin A). Xanthophylls in chicken and fish feeds. Phycobilins as food color, in diagnostics, cosmetics and analytical reagents.
VI. Source of fine chemicals	Glycerol used in foods, beverages, cosmetics, pharmaceuticals. Fatty acids, lipids, waxes, sterols, hydrocarbons, amino acids, enzymes, vitamins C and E. Polysaccharides as gums, viscocifiers and ion exchangers.
VII. Fuel	Long-chain hydrocarbons and esterified lipids as combustible oil. Hydrogen, biogas.
VIII. Hormones	Auxins, gibberllins and cytokines.
IX. Others	Biofertilzer, soil conditioner. Waste treatment.

enzymes, antibiotics, pharmaceuticals and several other fine chemicals (or their precursors), as well as hydrogen, hydrocarbons and other biofuels (methane, alcohol). These projected applications are summarized in Table 1.1.

Furthermore, the past years have witnessed the use of algae cultures for the assay of biological and physiological studies and proved a convenient medium for testing the effects of various chemical agents on living organisms.

At present, however, commercial algae production is still restricted to very few plants producing high-value health food or pigments, most located in the Far East and the USA. Health food production systems mainly cultivate the filamentous cyanobacterium *Spirulina* and the unicellular green alga *Chlorella* and – to a lesser extent – the green alga *Scenedesmus*, whereas for the production of pigments *Dunaliella* and *Spirulina* are preferred. In basic research, algae, by their structural simplicity and functional complexity, have become the vehicles for important discoveries and experiments. As a result, the cultivation of algae as a research tool is expanding rapidly as modern research becomes interested in such diverse fields as taxonomy, morphology, physiology, biochemistry, genetic engineering, tissue culture, food production, agriculture, waste disposal and medicine.

During the 1970s urgent efforts were made in finding alternative energy sources in view of the diminishing and ever more expensive supply of petroleum products. Thus the biological application of solar energy has attracted global interest. As algae constitute an efficient system for harnessing solar energy, there is continued interest in the technology of algae production as a source of renewable energy. Coupled with this development is the increased awareness of environmental pollution and waste disposal hazards. In this context, again, algae have a vital role to play in transforming sewage and waste water into valuable biomass.

As a result of the problems encountered with the popularization of microalgae as food or feed, different approaches for overcoming economic difficulties of algal production gained increasing attention. Instead of producing proteinaceous commodities of low commercial value, the priority of research was the utilization of high-value algal constituents, metabolites and fermentation products.

Although there are at present only a few chemicals isolated from microalgae that have gained commercial success, there is still a large number of algae species that are waiting to be screened for their ability to produce specific commodities.

Interest in commercial production of microalgae is not lacking and their economic potential is becoming widely recognized, yet little industrial effort in the mass production of microalgae is presently evident, several decades after the first large-scale algal ponds were constructed. In other words, a discrepancy clearly exists between the well-recognized commercial potential of many microalgal species and the actual industrial development. Two of the major reasons for the slow progress in industrial algal culture are the prohibitively high cost of production, which to a large extent is due to the relatively low areal outputs, and the acceptability difficulties linked with the popularization of algal biomass in any form.

Although some of the problems and critical factors encountered and identified during the 1950s could have been solved, it cannot be denied that several of them still await their solution and that several new problems have come up. Even at that time it was realized that the various problems associated with large-scale algae production are beyond microbiology and require co-operation with other scientific disciplines such as engineering, economy, medicine, toxicology, analytical chemistry, etc.

In this book, any economics for the production of algae or algal constituents are deliberately omitted. During the past 20 years, too many impressive calculations have been published based on extrapolations of algal yields obtained in small experimental ponds to large plants of several hectares, using costs of harvesting and processing techniques that have never been practically tested, and steps for the isolation of algal constituents or products – which presently are produced much more cheaply by other processes – leaving out no detail (including laser-guided levelling of the site up to the salary of the general manager). None of these designs and calculations have ever been fully or partly realized. They simply reflect the wealth of ideas and are no more than castles in the air. Algal biotechnology, however, is more than a playground for would-be ecologists or entrepreneurs. These futuristic calculations do not help to propagate algal biotechnology; on the contrary, they do harm to this field of research. Perhaps all these attempts may be excused by the fascination which emanates from the algae and the wishful thinking to 'domesticate' these microorganisms.

This book is not intended as a scientific review – although sufficient references are cited for almost any interested reader – but envisages the compilation of the major results published so far on various aspects of applied phycology. It aims at presenting the available information on the methods and applications of algal cultures in a simple and concise form, having in mind the interests of students and working microbiologists.

The enormous expansion of research using algal cultures and the constant development of new techniques and devices make it futile to claim any comprehensiveness. Effort has, however, been made to include the important references and also to give illustrative material as far as possible with the aim of providing a working guide on the various aspects of algal cultivation based on available information and on the author's own experience. It is hoped that the technical data presented here are of relevance and value for further development in the field of algal technology, and will generate additional interest in those who are committed to work on algae and their applications.

2 Algae production systems

For almost 50 years, numerous attempts have been made to exploit microalgae on a technological scale as a source of food, feed, lipids, vitamins, pigments, fertilizers, pharmaceuticals and other speciality chemicals. This common interest has arisen as a result of the need for additional food supplies, increasing problems of waste disposal and the shortage of raw materials and energy resources.

The cultivation of algae, especially under outdoor conditions, involves a very complex system, depending on the interaction of various external and internal factors. An overall, generalized schematic flow diagram of the various inputs and requirements involved in algae mass cultivation is given in Fig. 2.1. Although the individual parameters may differ from case to case, the basic elements will be similar in almost all cultivation units: an algal growth pond of varying shape and size which must receive all the necessary nutrients, harvesting and concentrating equipment, and a provision to process (extract, dry) the harvested algal biomass.

It has been stated repeatedly that algae cultivation is a special form of agriculture but is easier to handle and to manipulate than the conventional form. Indeed, at a first glance this judgement seems to be correct since some of the factors involved in the production of algae (such as fertilization, control of contamination or time of harvesting) can be operated more precisely. However, practice has shown that this is not true; there are many wealthy farmers but as far as we know nobody who became a rich man by the cultivation of algae (or writing about it). The several abandoned algal projects worldwide taught us that the successful growth of algae is 'high technology'; it is more or less an art and a daily tightrope act with the aim of keeping the necessary prerequisites and the various unpredictable events involved in algal mass cultivation in a sort of balance (Fig. 2.2). If at all, algae cultivation may be considered as an intermediate between that of agriculture and fermentation technology. Although the contrary has been reported repeatedly, it seems not to be true that microalgae have higher photosynthetic efficiencies than conventional crop plants. Nevertheless, algae production units can be more productive in terms of product output than conventional agriculture.

The research on applied algology in the past has led to the design and construction of several types of cultivation plants capable of producing quantities of algae ranging from kilograms to tons. Two main approaches

5

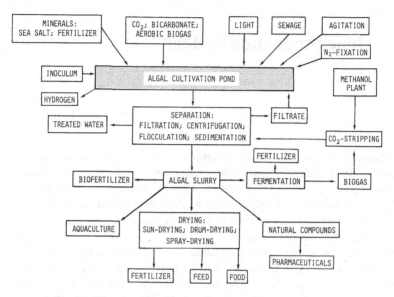

Fig. 2.1. Flow diagram of algae mass-cultivation system.

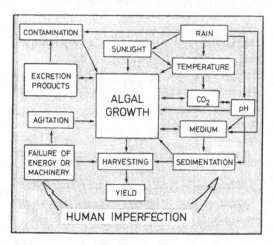

Fig. 2.2. Schematic diagram of the main interactions of algae cultures growing on sewage.

to the design of such plants have emerged from this work. They have been dictated primarily by the nature of the desired end-product, namely plants for the production of protein or food additives, which must be cheap to build and easy to maintain, and therefore suitable for developing countries, in order to compete effectively with other modes of food production; and more sophisticated biotechnologies with a high input of equipment and energy for the growth of specialized algal strains intended for the produc-

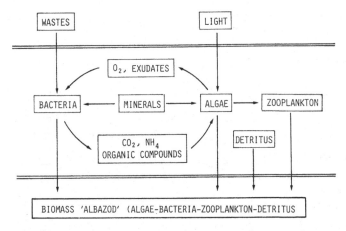

Fig. 2.3. Interaction of main factors influencing outdoor algae cultivation.

tion of specific biochemicals such as enzymes, pigments, therapeutic agents, etc. In such cases, the product quality justifies the large initial investment costs. For these purposes, three different systems can be distinguished, depending on the raw materials used and the utilization of the obtained algal biomass.

a) Systems in which a selected algal strain is grown in a so-called clean process, using fresh water, mineral nutrients and additional carbon sources. The algae in such systems are intended to be utilized mainly as food supplement.

b) Systems using sewage or industrial waste waters as the culture medium without the addition of mineral and external carbon. In such units the algal population consists of several species in the presence of high amounts of bacteria. The plant operation is based on a symbiotic nutritional circuit between algae and bacteria; i.e. the oxygen, photosynthetically produced and liberated by the algae, is utilized by the bacteria for aerobic biodegradation of organic compounds. During this process, organic carbon compounds are partially oxidized to CO_2, which in turn is assimilated by the algae. In addition, the algae utilize soluble nitrogen and phosphorus compounds from the medium and convert them into biomass, whereby more N and P can be consumed than are absolutely necessary for cell metabolism ('luxury metabolism') thus enhancing the waste water treatment (Fig. 2.3). Under limited or insufficient light conditions, certain algae adapt their metabolism from autotrophy to heterotrophy so that they are able to utilize both inorganic and organic compounds from the medium. The biomass thus produced consists of a mixture of algae, bacteria and zooplankton.

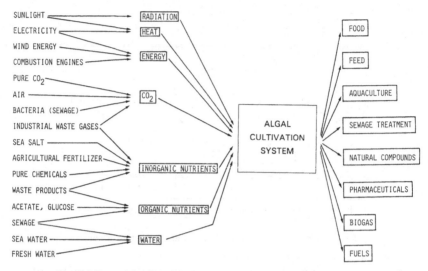

Fig. 2.4. Potentials of algal biomass: raw material, energy sources, products and applications.

c) Cultivation of algae in enclosed systems under sunlight or artificial light, with cells being grown preferably in autotrophic media.

It is obvious that algae production plants for mass cultivation have to be located in areas with suitable climatic conditions. Unfortunately, places where all the involved parameters are optimal are very rare. Some locations which offer favourable climatic conditions have to be excluded because of the absence of the required infrastructure.

In the early work of algal mass cultivation, a variety of different algae were investigated for their growth characteristics, particularly for fast multiplication rate and high protein content. The species that are suitable for this purpose in the 'clean' system are *Chlorella, Scenedesmus, Uronema, Coelastrum, Dunaliella, Porphyridium* and *Spirulina*; and in sewage or waste water, *Micractinium, Oocystis, Oscillatoria, Chlamydomonas, Euglena* and *Ankistrodesmus*. Among these, the most common species used nowadays for mass cultivation are *Chlorella* with its strains *C. pyrenoidosa, C. ellipsoidae* and *C. vulgaris, Scenedesmus* with its strains *S. obliquus (acutus)* and *S. quadricauda, Dunaliella bardawil, Porphyridium cruentum*, and the only cyanobacterium *Spirulina*, with its not always clearly defined strains *S. maxima, S. platensis* and *S. geitleri*. Almost all of the results, data and experiences described here are based on studies performed with these algae.

The various different potential inputs and applications that have emerged from the past research activities on the production and utilization of microalgae are summarized in a simplified form in Fig. 2.4.

3   Culture media

Under natural conditions, most algae grow as mixed communities, which include various species and genera. In order to study and cultivate individual species, the required form must be cultivated without other species present. The isolation of the desired species depends on the provision of a suitable environment for its growth. Forms of algae which grow more rapidly under the conditions provided will outnumber the other species after a certain period of time. If not replaced or fortified, the composition of the medium will gradually change thus limiting the growth of the originally predominant species, and other species, also introduced into the medium, will come up. If such cultures are maintained for longer periods of time, successive development of many different algal species will take place in a single culture.

For most of the ecological and physiological studies performed nowa-days with algae, pure algal cultures are a basic requirement. Such cultures have greatly contributed to many physiological discoveries and are still being used in various morphological, biochemical and genetic researches. Many studies on antibiotic resistance, variability, teratology, cytology etc., have been made possible with the help of authentic algal cultures. Several terms can still be found in the literature for pure cultures, namely unispecific, unialgal or axenic culture. Whereas the first two terms refer to cultures which contain only a single species of alga in the presence of other organisms, the latter is used for cultures which contain only one species of alga, free from any other organism.

The development of synchronous cultures is another important example of the usefulness of culture methods and has facilitated the understanding of various biochemical reactions at cellular level. Since the early 1940s, when the possibility of utilizing fast-growing unicellular algae as a source of food or feed was suggested, algal cultures have assumed an industrial dimension.

For the successful growth of algae in a culture, the environment must be conditioned to meet as many of the intrinsic requirements of that organism as possible. The environmental factors may be either physical, such as temperature and light, or chemical, which provides all the raw materials used for the structural and protoplasmic synthesis of the algal cell.

Studies on the nutritional requirements of algae, using the replacement technique as employed for higher plants, have revealed that the requirements of algae are similar to those for the phanerogams. In 1896, Molisch had observed that the mineral nutrition of the algae was not different from that of higher plants. The major absolute requirements include carbon, phosphorus, nitrogen, sulfur, potassium and magnesium. Elements like iron and manganese are required in small amounts; sodium is not essential for many forms and the specific requirement for calcium is contradictory. Various other elements like cobalt, zinc, boron, copper and molybdenum are essential trace elements. In addition to these basic minerals, several algae (heterotrophs, see below) require additional organic substrates (vitamins, nucleic acids, growth factors) for their growth.

Two major forms of algal nutrition can be distinguished, i.e. autotrophy and heterotrophy. Autotrophs are all those organisms that obtain all the elements they require for growth from inorganic compounds only, whereas heterotrophs are organisms that need organic substrates synthesized by other organisms. The term amphitrophy is used sometimes to describe organisms able to live either auto- or heterotrophically.

Among the autotrophs two groups can be classified, namely photoautotrophs and heteroautotrophs. Photoautotrophs are those organisms that obtain the energy for their metabolism from light; chemoautotrophs obtain this energy from the oxidation of inorganic compounds or ions. Most algae are phototrophs; they obtain their energy through the absorption of light and electrons used for the reduction of CO_2 by the oxidation of inorganic substrates, predominantly water, coupled with the evolution of oxygen. If the organisms cannot grow in the dark, which is the case with most algae, they are called obligate phototrophs. Several intermediate variations can be found among the heterotrophs. For organisms whose energy is derived from the oxidation of organic compounds that also serve as a carbon source, the term chemoheterotrophy is used.

Photoheterotrophy is the term for those organisms that require light as an energy source to utilize organic carbon. Auxotrophy is a form of heterotrophy in which organisms require only very small quantities of essential organic compounds (vitamins, amino acids, etc.). Mixotrophy designates a nutritional mode in which the energy is derived either through photosynthesis or by chemical oxidation, but CO_2 and organic carbon sources are necessary to sustain growth.

Phagotrophy describes the uptake of solid organic particles by the organism (phagocytosis) and digestion in food vacuoles. Many media have been devised for culturing different algae, but no single medium can be said to be the best one. The main considerations in developing nutrient recipes for algal cultivation can be summarized as follows (Vonshak, 1986).

a) Total salt concentration, dependent on the original biotope of the alga.

b) Carbon source. Since about 50% of the algal biomass consists of carbon, a sufficient supply of carbon is of vital importance for successful cultivation. Carbon can be supplied as an inorganic substrate in the form of gaseous CO_2, as is the case for most photoautotrophic forms, or in the form of bicarbonate. Organic carbon sources are mainly sugars or acetate.

c) Choice of a suitable, and also economic, nitrogen source. Nitrate, ammonia and urea are widely used, dependent on the species and the pH optimum. Besides carbon, nitrogen is the most important element in algal growth since more than 10% of the algal biomass may consist of nitrogen. In addition, changes in the nitrogen supply essentially influence the metabolic pathways in the alga and subsequently the overall composition of the organism.

d) Concentration of the other major elements such as potassium, magnesium, sodium, sulfate and phosphate.

e) The pH value of the medium. Usually neutral or slightly acidic values are used, mainly to avoid precipitation of several major elements.

f) Trace elements, which are found to be essential and which are mostly supplied in very small quantities from stock solutions. To enhance the solubility of some of these elements (especially iron), chelating agents such as EDTA are used.

g) Addition of organic components and growth promoting substances (vitamins, hormones, etc.).

Many kinds of culture media have been formulated to suit the requirements of the various algae and the idiosyncrasies of the scientists. To be of value, the medium must support the species to be cultivated by supplying the necessary minerals or organic compounds ('growth factors') and energy sources required for growth and cell synthesis. Unfortunately, too many cooks have tried to modify already well-established recipes for media so that a considerable amount of confusion exists in the nomenclature. A proper citation of the source of the culture medium is indispensable; but it is not unusual that authors fail even to mention whether the salts used are anhydrous or not.

The culture media that are employed for algae can be broadly grouped into three major categories, namely a) complete synthetic media, b) those that are based on natural waters enriched by mineral supplementation, and c) liquid wastes such as effluents from fermented wastes or from waste water treatment plants, industries, etc. Whereas some of the media are designed to simulate natural conditions as closely as possible, others are formulated to obtain optimal growth rates.

CO_2 as carbon source

Like any other autotrophically growing plants, algae require an inorganic carbon source to perform photosynthesis. Since algal cultures, in contrast

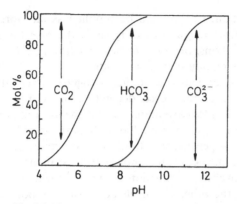

Fig. 3.1. The pH-dependency of $CO_2/HCO_3^-/CO_3^{2-}$ equilibrium.

to conventional agriculture, cannot be supplied with carbon by simple diffusion from the air – the natural CO_2 concentration in air (0.03%) is too low to sustain optimal growth and high productivity – cultures of algae, growing in fresh water with low salinity and at nearly neutral pH, must be supplied with additional carbon to ensure satisfactory growth. Although intensive mixing may increase the entrance coefficient for CO_2 from the air into the culture, in most of the algal mass cultures, additional carbon – mainly in the form of CO_2-enriched air – is supplied. An exception is *Spirulina*, which is able to utilize carbonate or bicarbonate, supplied in the form of salts.

In water, CO_2 may appear in the following forms:

$$CO_2 + H_2O \longleftrightarrow H_2CO_3 \longleftrightarrow H^+ + HCO_3^- \longleftrightarrow 2H^+ + CO_3^{2-}$$

depending on the pH, the temperature and the concentration of the nutrients in the medium (Fig. 3.1). In a poorly buffered system, as it is the case with many algae culture media, the assimilation of carbon dioxide or bicarbonate by the rapidly growing algae causes a shift of the above equilibrium resulting in elevated pH values due to the excretion of OH^- ions by the algae into the medium. In mass cultures, the pH must be maintained in order to keep it in the optimum range for the cultivated species and to prevent the depletion of carbon in the medium.

It has to be remembered that the rates of the principal chemical reactions that involve the solubility of carbon in water are quite different. Whereas the ionization reactions

$$CO_3^{2-} + H^+ \longleftrightarrow HCO_3^- \text{ and } HCO_3^- + H^+ \longleftrightarrow H_2CO_3$$

are almost instantaneous, the following reaction

$$H_2CO_3 \longrightarrow H_2O + CO_2$$

is a relatively slow process with $k(H_2CO_3) = 26.6 \text{ s}^{-1}$

Chemical analyses have shown that algal biomass consists of about 50% carbon, which means that about 1.8 kg of carbon dioxide are required to produce 1 kg of algae. Since pure carbon dioxide is expensive, alternative and cheaper sources of this gas, such as waste gas from industrial combustion processes, diesel engines, cement plants or fermentation have been tried. Ethanol production as the main agro-industrial fermentation process is considered as an ideal source of carbon dioxide for algal cultivation as the gas can be used without costly purification. Therefore, the location of algae ponds near such plants is an important consideration in reducing algae production costs; so far, however, all these possibilities remain unrealized and have not gained any economical importance. The current prices for commercial delivered CO_2 are in the range of US\$ 40–60 t^{-1}. Algal systems suffer the disadvantage that CO_2 transfer from air can only occur on the single surface of the air/water interface, unlike higher plants that maximize this interface with a greater leaf area.

Organic carbon sources

The nutrient requirements of microalgae, even within the same species, differ widely, the pattern ranging from obligate autotrophy to obligate heterotrophy. In several algal species, the mode of carbon nutrition can be shifted from autotrophy to heterotrophy by modifying the carbon source, based on the ability of the naturally photoautotrophic algae to utilize both inorganic and organic carbon substrates. This flexibility in carbon uptake has been studied in detail, namely with the two chlorophyceae *Scenedesmus* sp. and *Chlorella* sp., both important species for large-scale production. It was found that these algae can shift between autotrophy and heterotrohy in the sense that they are capable of growing on CO_2 in light and on certain forms of organic carbon during darkness. It can be assumed that the lack of versatility in using several sugars (or other organic carbon substrates) is due to limitations in cell-wall permeability, because several sugars, which, added to the medium without considerable growth-enhancing effect, can be found as naturally occurring intermediary metabolic products within the algal cells. The growth-stimulating effect of different sugars is pronounced during light-limiting conditions or in the dark, whereas under saturated light no additional increase in the growth rate can be observed.

On the other hand, it was reported that *Dunaliella* is apparently unable to grow heterotrophically, i.e. utilizing organic carbon as the sole source of carbon (Ben-Amotz & Avron, 1989). It was described that it was impossible to grow this alga with acetate or glucose in the night and it was concluded that this organism is obligate photoautotrophic.

For the development of an economic algal mass cultivation, especially in developing countries, where the use of CO_2 is almost prohibited due to the high costs, it is necessary to explore the possibilities of utilizing low cost carbon sources such as industrial or agricultural by-products.

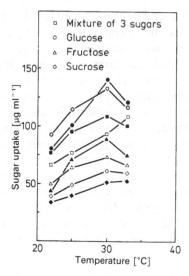

Fig. 3.2. Uptake of different sugars by *Scenedesmus obliquus* during dark and light phase at various temperatures. (Open symbols = light, closed symbols = dark.)

Sugars and molasses

One of the potential sources of organic carbon is molasses, which is available in many sugar-cane growing countries.

To study the possibilities of enhancing the growth of *Scenedesmus obliquus* and reducing the amount of CO_2 needed, several trials were performed in India by either adding different sugars or molasses to the medium. To evaluate growth stimulating effects of the individual sugars on laboratory cultures, 0.25 g 1^{-1} of glucose, fructose, sucrose, mannose and galactose were added as carbon substrate. The quantity of sugar absorbed in the medium was determined and the initial concentration was made up daily by adding the corresponding amount. No other carbon source was provided except the CO_2 present in compressed air supplied at a rate of 300 ml min^{-1} 1^{-1} algal culture. The growth-promoting effect of the different sugars tested and the influence of different temperatures on the amount of sugar absorbed is shown in Figs 3.2 and 3.3. The highest increase in biomass was obtained with glucose, followed closely by mannose, whereas the other sugars resulted in significantly lower growth rates.

A cheap source of glucose is molasses, which on an average contains about 25% glucose, 25% fructose and 30% sucrose. To test the feasibility of using commercial sugar-cane molasses, series of trials were performed with outdoor cultures of *Scenedesmus obliquus* by adding various concentrations of molasses (50, 100, 150 and 300 mg 1^{-1}) to the medium. The aim of these tests was to determine the maximum growth-enhancing effect and

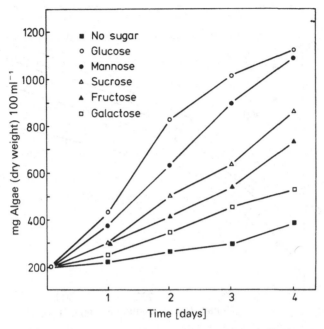

Fig. 3.3. Growth of *Scanedesmus obliquus* after addition of various sugars (25 mg ml^{-1}, 10 h light, 14 h dark; addition of sugars at end of light phase).

the maximum concentration of molasses that can be absorbed completely by the algal cells during the night so that a period of heterotrophic growth (night) is followed by a period of autotrophic growth (day). It was expected that by this schedule the bacterial contamination in the culture could be maintained at a low level, since during the day the amount of organic substrate is low in the medium, which prevents the bacteria from entering into a prolonged logarithmic growth phase. It was found that with algal concentrations between 200 and 500 mg l^{-1} a substantial increase in growth was obtained with molasses concentrations up to about 150 mg l^{-1}. Higher molasses concentrations did not significantly further increase the growth rate (Fig. 3.4). On the other hand, it was observed that only concentrations less than 140 mg l^{-1} of molasses (corresponding to about 100 mg of total sugars per liter) could be absorbed completely during the night. At this concentration the bacterial counts in the cultures did not rise substantially as compared with controls grown without the addition of molasses.

The beneficial effect of molasses as a supplementary carbon source could be demonstrated by a number of outdoor cultivation experiments. Maximum algal growth was obtained under conditions when molasses and CO_2 were supplied in combination, i.e. CO_2 during the day and molasses during the night (Fig. 3.5). Major differences in the protein content between the samples grown under the various conditions could be detected. As can be

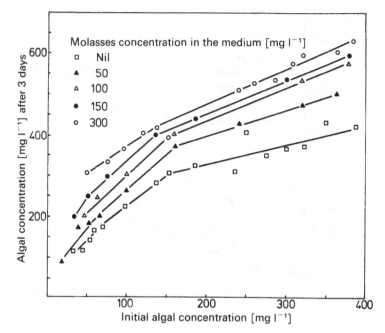

Fig. 3.4. Effect of various molasses concentrations on the growth of *Scenedesmus obliquus*.

seen from Table 3.1, highest protein concentrations were obtained in the cultures grown with a combination of air, CO_2 and molasses (Mode D). It should be stressed once more that the concentration of protein was highest in the samples harvested in the morning and lowest in samples harvested in the evening. The relevance of this fact has to be considered when schedules for pond operations are established. It might even happen that under sub-optimal conditions the total amount of protein harvested in the evening is smaller than it would have been in the morning of the same day, in spite of an substantial increase of biomass during the course of the day. Besides protein, the proportion of the other cell constituents, especially ash, also varies depending on the carbon source (Table 3.2). It is a remarkable observation that in samples grown without additional CO_2, high ash contents exceeding 20% were analysed. It has to be assumed that, because of the slow multiplication rate, the individual algal cells accumulate higher amounts of minerals from the medium resulting in the elevated ash contents.

Acetic acid
The utilization of acetic acid as organic carbon source is or has been practised in several algal plants, especially in those that are producing *Chlorella* as a health food. Theoretically, 1.25 l of acetic acid are required

Table 3.1. *Effect of carbon source and harvesting time on the protein content (% of dry matter) and yield per area of* Scenedesmus obliquus *(11 m², 1300 l)*

Harvesting time	Parameter	Mode of carbon supply[a]			
		A	B	C	D
Second day of cultivation					
End of dark period	% protein	42.5	43.0	40.0	58.2
	Protein (g m⁻²)	37.0	44.0	41.0	88.0
End of light period	% Protein	19.0	22.2	26.0	37.0
	Protein (g m⁻²)	33.0	46.0	55.0	77.0
Third day of cultivation					
End of dark period	% Protein	37.0	31.5	37.6	39.0
	Protein (g m⁻²)	64.0	72.0	86.0	120.0
End of light period	% Protein	19.0	25.0	26.0	37.0
	Protein (g m⁻²)	43.0	84.0	87.0	140.0

Note:
[a] Same conditions as described in Table 3.2.

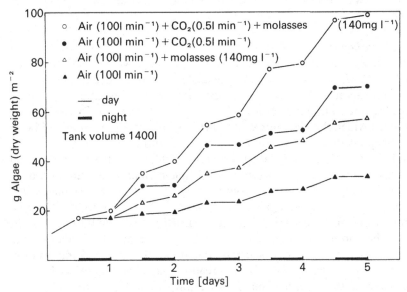

Fig. 3.5. Growth of *Scenedesmus obliquus* under different autotrophic and mixotrophic culture conditions.

for the production of 1 kg of algal biomass (= 0.5 kg carbon). In actual production, only 50–60% of the acetic acid is utilized by the algae so that about 2.0–2.5 l of acetic acid are needed per kilogram of algal biomass. High prices for acetic acid and difficulties in adjusting the pH within the

Table 3.2. *Effect of different carbon sources on gross chemical composition (% of dry matter) of* Scenedesmus obliquus *grown in outdoor ponds (11 m², 1300 l)*

Constituent	Composition of medium[a]			
	A	B	C	D
Crude protein	41–46	47–51	52–59	52–62
Carbohydrates	6–8	7–8	7–8	7–8
Crude fibre	9–17	10–12	12–14	9–14
Lipids	8–11	8–10	10–13	9–11
Nucleic acids	2–4	2–7	4–7	5–7
Ash	18–24	19–23	7–15	7–15

Notes:
[a] Basis medium plus: A, 100 l air min^{-1}; B, 100 l air min^{-1} + 100 mg total molasses sugar l^{-1} d^{-1}; C, 100 l air min^{-1} + 30 l CO$_2$ h^{-1} for 8 hours d^{-1}; D, air min^{-1} + 30 l CO$_2$ h^{-1} for 8 hours d^{-1} + 100 mg molasses sugar l^{-1} d^{-1}.

optimal range restrict the application of this carbon source to special cases only.

Hydrocarbons
Since it was reported that certain cyanobacteria grow in petroleum mixed with water (Enebo, 1967), a variety of algae have been tested for their ability to utilize alkanes. Algae were screened on the basis of metabolizing at least one organic compound. Among the algae tested, only two forms could be isolated which showed improved growth in a medium containing 0.1% (v/v) *n*-heptadecane (Payer, 1977). These were cultures of *Chlorella vulgaris* and *C. ellipsoidea*. Some algal samples isolated from the grounds of oil refineries (*Scenedesmus brevicauda*, *S. quadricauda*) showed improved growth of about 20% in media containing up to 1% (v/v) *n*-heptadecane. However, none of the algae isolated showed any ability to grow heterotrophically on hydrocarbons alone. Marked growth inhibition resulted with the addition of alkanes of 14 carbon atoms or less. Studies on the growth of algae on hydrocarbons might be of ecological importance, particularly because of the wide spread of oil spills, but so far have no relevance in the context of mass cultivation.

Nitrogen
Next to carbon, nitrogen is quantitatively the most important element in algal nutrition. An essential factor of the preparation of culture media for many algae is the form in which the nitrogen is supplied. Generally, algae are able to utilize nitrate, ammonia or other organic sources of nitrogen

such as urea. Nitrite can also be used, but its toxicity at higher concentrations makes it less convenient. In addition, certain cyanobacteria are also capable of assimilating nitrogen in its elemental form from the atmosphere. In practice, the preferred nitrogen supply is in the form of ammonia or urea, either of which is economically more favourable than nitrate or nitrite, which is more expensive and requires considerable metabolic energy for its assimilation, because the latter must be reduced before either can be incorporated into organic molecules. However, it has been reported that ammonia concentrations exceeding 5 mM, especially in combination with pH values above 8 are toxic for *Dunaliella*.

The growth rate is generally the same on both nitrogen sources. The average nitrogen requirement for many green algae is approximately 5–10% of the dry weight or 5–50 mM. However, there are considerable variations with this element, since the nitrogen content in the medium can be manipulated in such a way as to produce nitrogen-deficient cells or cells showing luxury consumption of this element.

Nitrogen content is also low in diatoms, where the silica of the cell wall substantially contributes to the dry matter, and in nitrogen-deficient algae that have accumulated large amounts of carbon compounds, i.e. polysaccharides or lipids. A low nitrogen supply may result in a low respiration rate and an increase in the lipid reserves in several algae. Excessive nitrogen absorption can be induced if the algae are grown in media deficient in manganese, boron, or zinc.

It was generally assumed that, before assimulation, nitrate-N is reduced to ammonium-N in four reducing steps, each adding a pair of electrons:

$$HNO_3 \xrightarrow[2e^-]{} HNO_2 \xrightarrow[2e^-]{} H_2N_2O_2 \xrightarrow[2e^-]{} NH_2OH \xrightarrow[2e^-]{} NH_3$$

More recent opinions suggest that the entire reaction is catalyzed by two enzymes in the following pattern:

$$NO_3^- \xrightarrow[2e^-]{} NO_2^- \xrightarrow[6e^-]{} NH_4^+$$

where the enzyme complex nitrate reductase/nitrate oxidoreductase catalyzes the reduction from nitrate to nitrite and the complex nitrite reductase/nitrate oxidoreductase in a direct reaction requiring six electrons as the step to ammonium.

Two different types of nitrate reductase have been isolated so far. One is in eucaryotic algae, where the enzyme is similar to the one found in higher plants, containing molybdenium, heme and flavin adenine dinucleotide. The second type is found in cyanobacteria, associated with chlorophyll-containing particles, lacking flavin and cytochrome and is using ferredoxin as an electron donor.

In artificial algae cultures, where nitrogen is usually supplied as nitrates or as ammonium salts, most algae can assimilate either, but evidence

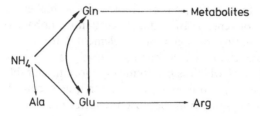

Fig. 3.6. Pathway of ammonia metabolism.

suggests that the ammonical nitrogen is preferently absorbed by many forms and nitrate is often not utilized until all the ammonium salts have been consumed. The addition of ammonium to algal cultures assimilating nitrate causes immediate and complete inhibition of nitrate assimilation. Since ammonia is the end product of nitrate reduction, it causes feedback inhibition and repression of nitrate uptake. Ammonia is converted to organic nitrogen compounds and its assimilation takes place at the expense of endogenous carbohydrate reserves. Glutamic acid is the first compound synthesized by the assimilation of ammonia, formed from its precursor α-glutaric acid and catalyzed by the enzyme glutamine dehydrogenase. Another pathway of ammonia assimilation has been described for cyanobacteria, where glutamine is the first product. The main pathways of ammonia assimilation are summarized in Fig. 3.6. Supply of ammonium to N_2-fixing cyanobacteria inhibits the activity of nitrogenase and cell differentiation to heterocysts.

The assimilation of nitrate and ammonia is closely related to the pH of the medium, since nitrogen absorption changes the pH. When ammonia is used as the sole nitrogen source, the pH of the medium may decrease rapidly to as low a level as pH 3.0, causing deleterious side effects. It has been reported repeatedly that brief periods of high ammonia concentration can be a useful tool in eliminating contamination by herbivorous zooplankton. However, certain algae are sensitive to high ammonium concentrations and their growth might be inhibited by concentration about 1 mM. Assimilation of nitrate ions tends to raise the pH. These pH shifts might be the reason for the growth inhibition observed with some algae caused by elevated ammonium concentrations in the medium, i.e. an increase in intracellular pH due to the penetration of undissociated ammonium hydroxide molecules. Ammonia will have the tendency to volatilize from the medium; however, several microalgae can take up ammonia in excess of the immediate metabolic needs (luxury consumption) so that the supply can be adjusted to other environmental conditions such as pH or temperature.

Although it is well established that uptake and assimilation of inorganic nitrogen is stimulated by light in several ways, the mechanism of this light stimulation is not yet fully understood. One possibility is the supply of ATP by photophosphorylation, supported by the observation that uncoupling of photophosphorylation inhibits nitrite and nitrate uptake. It seems that ATP is mainly required for the active intracellular uptake of nitrate rather than for its reduction.

The second way by which light reacts on N_2 metabolism is the direct photoreduction of the N-compound by NAD(P)H or ferredoxin, both derived from photosynthetic electron transport. It has been reported that it is mainly the nitrite reducing step that is affected, as it depends on ferredoxin and the continuous supply of reductants originating from photosynthesis.

The third possibility of light-influenced nitrogen uptake is the photosynthetic production of carbon skeletons that couple to the reduced nitrogen, confirmed by the requirement of light and CO_2 for the assimilation of any form of nitrogen. In certain algae a nitrogen-uptake limiting deficiency of carbon molecules can be substituted by organic carbon sources such as acetate, glucose or storage polysaccharides.

Among the organic nitrogen sources, urea is the one which has gained importance especially in large-scale algal cultivation, because it is a good potential nitrogen source for almost all algae. Urea can serve as the sole nitrogen source for several unicellular members of *Chlorophyta*, *Xanthophyta* and cyanobacteria, and it has been used as the best nitrogen source for the mass cultivation of *Chlorella* and *Scenedesmus*. Its usefulness in *Spirulina* cultures is not yet established because contradictory results have been reported, which might be caused by different culture conditions, especially pH and bicarbonate concentration.

It is assumed that urea is hydrolyzed before the nitrogen is incorporated into the algal cells by two different possible pathways.

1) Urease catalyzed hydrolytic reaction:
$$CO(NH_2)_2 + H_2O \longrightarrow 2NH_3 + CO_2$$

2) Two-step reaction with the formation of allophanate as an intermediate:

   a) catalyzed by urea carboxylase

$$CO(NH_2)_2 + HCO_3^- \longrightarrow O=C\begin{smallmatrix} \nearrow NH-COO^- \\ \searrow NH_2 \end{smallmatrix} + H_2O$$

   b) catalyzed by allophanate amidohydrolase

$$O=C\begin{smallmatrix} \nearrow HN-COO^- \\ \searrow NH_2 \end{smallmatrix} + 2H_2O + OH^- \longrightarrow 2HCO_3^- + 2NH_3.$$

It is accepted that algae able to assimilate urea possess either the first or the second enzymatic system, but not both. The effect of transferring exponentially growing algal cells to a medium lacking a source of assimilable nitrogen is of interest from both the physiological and ecological points of view. In algae treated this way, the flow of carbon fixed during photosynthesis is switched from the path of protein synthesis to that leading to carbohydrate; cell division does not continue after the amount of cell nitrogen has fallen below a limited value. On prolonged nitrogen starvation, lipids begin to accumulate in the cells as a result of the lipid-synthesizing enzymes being less susceptible to disorganization than the enzymes responsible for the carbohydrate synthesis, so that the major proportion of carbon is bound in lipids. This method of forcing algal cells to synthesize larger proportions of lipids was suggested as a possibility of meeting the lipid shortage in Germany during World War II. Another application of manipulating the chemical composition of algae by modifying the growth conditions is the production of glycerol and β-carotene by the halotolerant green alga *Dunaliella*.

Certain cyanobacteria possess the ability to utilize nitrogen storage compounds. It has been reported for *Spirulina* and *Anabaena* that the accessory pigment phycocyanin is degraded during nitrogen starvation and is freshly synthesized when nitrogen becomes available.

One of the major drawbacks of algal cultures grown under nitrogen limitation is the fact that although some valuable products may be synthesized and accumulated, the overall productivity is significantly reduced. That means that calculations predicting high yields of lipids must be tempered by the fact that the time required to obtain the predicted amounts is longer than the time needed to yield the same amount from non-starved algae.

The situation looks more promising for the production of carbohydrates by algae grown under nitrogen-limited conditions. It has been shown that several algae (*Scenedesmus* sp. and *Spirulina* sp.) are able to produce high concentrations (about 70%) of carbohydrate polymers without a decrease in overall productivity. The effects of different nitrogen regimes on growth parameters and constituents have been studied for the two chlorophyceae *Scenedesmus obliquus* and *Chlorella vulgaris*, and for four cyanobacteria (*Spirulina platensis, Anacystis nidulans, Microcystis aeruginosa* and *Oscillatoria rubescens*). The major results are summarized in Table 3.3. In all algae, increasing nitrogen levels lead to an increase in biomass, protein content and chlorophyll. At low nitrogen levels, the chlorophyceae contained high levels of total lipids (45%), more than 70% of which were neutral lipids containing mainly 16:0 and 18:1 fatty acids. At high nitrogen levels the percentage of total lipids dropped to about 20% with polyunsaturated fatty acids predominating. The cyanobacteria, however, did not show significant changes in their fatty acid and lipid composition under varying nitrogen concentrations in the medium. Thus, the chlorophyceae, but not the cyanobacteria, can be manipulated in mass cultures to yield a biomass with desired lipid composition.

As already mentioned above, some of the cyanobacteria are capable of fixing atmospheric nitrogen by the reduction of N_2 to NH_4^+, catalyzed by the enzyme nitrogenase. About 30 different cyanobacteria have been

Table 3.3. *Effect of different nitrogen concentrations on the chemical composition of various algae*

	N-source	0.0003%	0.001%	0.003%	0.01%	0.03%	0.1%
A. Total protein (% of dry weight)							
Chlorella vulgaris	NH_4Cl	7.79	11.1	19.9	28.9	31.2	—
Scenedesmus obliquus	NH_4Cl	9.36	9.43	22.0	33.2	34.4	—
Chlorella vulgaris	KNO_3	12.6	6.75	14.5	30.7	31.1	32.2
Scenedesmus obliquus	KNO_3	8.19	9.00	8.81	34.0	32.1	32.9
Anacystis nidulans	KNO_3	21.2	18.3	33.4	33.9	39.7	46.3
Microcystis aeruginosa	KNO_3	28.1	27.6	23.5	24.9	46.5	50.1
Oscillatoria rubescens	KNO_3	*	*	28.0	35.6	53.8	48.6
Spirulina platensis	KNO_3	*	25.8	26.6	33.4	52.1	47.4
B. Total lipids (% of dry weight)							
Chlorella vulgaris	NH_4Cl	52.8	41.8	20.2	14.1	11.8	—
Scenedesmus obliquus	NH_4Cl	34.6	33.1	21.7	23.0	22.4	—
Chlorella vulgaris	KNO_3	57.9	62.9	42.7	22.0	21.8	22.6
Scenedesmus obliquus	KNO_3	45.6	44.3	50.1	26.9	29.8	21.2
Anacystis nidulans	KNO_3	14.3	12.7	13.6	12.0	15.4	14.8
Microcystis aeruginosa	KNO_3	17.7	13.7	13.7	13.8	18.2	23.4
Oscillatoria rubescens	KNO_3	*	*	12.7	12.0	11.3	12.8
Spirulina platensis	KNO_3	*	11.2	9.1	12.6	15.5	21.8

Note:
* The alga could not be grown at this nitrate concentration.
Source: After Piorrek *et al.*, 1984.

identified that are able to fix N_2, and it has to be assumed that there are many more species belonging to this group.

From the information available the general conclusions that can be drawn are 1) not all cyanobacteria are able to fix N_2; 2) N_2-fixation is not necessarily linked to the formation of heterocysts as N_2-fixation can also be performed, aside from heterocystous filamentous strains, by unicellular and filamentous strains lacking heterocysts; 3) the latter two groups can fix N_2 only at low oxygen levels; whereas 4) heterocystous algae can fix N_2 in the presence of atmospheric levels of oxygen. There is no doubt that the heterocyst is the site of aerobic N_2-fixation.

The best studied heterocystous forms of N_2-fixing cyanobacteria are probably *Anabaena* sp. and *Nostoc* sp. In the absence of bound forms of nitrogen, these algae differentiate vegetative cells into heterocysts which contain the enzyme nitrogenase. Nitrogenase is a complex enzyme, composed of a molybdenium–iron protein and a smaller iron protein, which both together catalyze the following reaction:

$$N_2 + 6H^+ + 6e^- + 12\ MgATP \longrightarrow 2NH_3 + 12\ MgADP + 12\ P_i.$$

This enzyme system is also able to catalyze a variety of other substrates, i.e. N_3^-, N_2O, HCN, CH_3CN, C_2H_2, etc. For these reactions it needs the presence of ATP and Mg^{2+} and a suitable reductant, mainly carbon compounds produced during photosynthesis. Many algae from different taxonomic groups are able to utilize various organic nitrogen compounds, besides urea, as nitrogen source, for instance amides, asparagine, glutamine and other amino acids, as could be confirmed with ^{14}C-labeled compounds.

Phosphorus

Phosphorus is a major nutrient required for normal growth of all algae; it is essential for almost all cellular processes, i.e. biosynthesis of nucleic acids, energy transfer, etc. The phosphorus concentration is often growth limiting in the natural aqueous habitat, where it occurs as orthophosphate and in organic combinations.

Algae vary among themselves in their ability to utilize organic phosphorus, even though they contain phosphates, and normally these compounds must be hydrolyzed by extracellular phosphatases. Pyrophosphate is not utilized as readily as orthophosphate. Under natural conditions, other organic phosphate compounds are broken down by bacteria, releasing phosphate.

The major form in which algae acquire phosphorus is as inorganic phosphate, either as $H_2PO_4^-$ or HPO_4^{2-}. The optimum phosphorus concentration in the medium, as well as the phosphorus tolerance, varies with different species, even if all other nutrients are supplied in sufficient concentrations. The average tolerance for most algae is in the range 50 μg l^{-1}–20 mg l^{-1}.

Incorporation of orthophosphate into the algal cells is an energy-dependent process, by which the energy is provided either by photosynthesis or by respiration, which predominantly proceeds in three different ways: 1) photophosphorylation; 2) substrate phosphorylation; and 3) oxidative phosphorylation. In the first process, which is common to all higher plants, the photosynthetic-produced energy is bound to the energy-rich phosphorus ATP which serves as the central energy carrier for all energy-requiring processes in the living cell. In the other two processes, the energy is derived from the respiratory chain in the mitochondria or from the oxidation of respiratory substrates. Under sufficient phosphorus supply, inorganic phosphates are accumulated in the algae as acid-labile polyphosphates in large granules, which are 'metabolized' under phosphorus deficiency.

However, the uptake of this nutrient is still affected by other factors such as pH and the concentration of Na^+, K^+, Mg^+ or various heavy metals in the medium. It has been reported that, similar to the effects observed with algae grown under nitrogen starvation, phosphate-deficient algae cultures tend to accumulate large amounts of lipids, while the content of protein, chlorophyll and nucleic acids decreases. Also, an accumulation of cyanophycin granules was observed in phosphate-deficient cells of some cyanobacteria. After starvation, the phosphorus uptake is much higher than in unstarved cultures.

Sulfur

Like phosphorus, sulfur is vital to all cells, because it is a constituent of some essential amino acids (methionine, cysteine, cystine), vitamins and sulfolipids, etc. It is generally provided as inorganic sulfate in the culture medium. Some algae (*Chlorella, Euglena*) have been reported to be able to utilize organic sulfur sources such as sulfur-containing amino acids under certain conditions. Sulfur uptake is an active, phosphorylation-driven, energy (light)-requiring and temperature-sensitive process. The need for energy and reductants, similar to the uptake of nitrogen, could be demonstrated by the use of photosynthesis uncouplers in both light and dark. Like nitrate, most of the sulfate taken up by the algae has to be reduced before it is incorporated into cellular material, but short-term sulfate uptake, independent of its subsequent reduction, has also been reported.

In the dark and under anaerobic sulfide-rich conditions, some algae are able to grow heterotrophically using sulfide as an electron donor with the liberation of sulfur. The largest part of the absorbed sulfur is incorporated into protein; a certain quantity of this element can also be found as component in sulfolipids and as sulfuric acid in the vacuolar sap of some algae.

Calcium

The absolute requirement for calcium for the growth of algae is in dispute, and for a long time this element was not thought to be an absolute requirement. Its physiological role is still not fully understood; however, it is accepted that where calcium is required, as for instance in green algae, the amount is small and often might have been supplied as an impurity with other nutrients. Calcium ions may play a role in the maintenance of cytoplasmic membranes, salt formation with colloids and the precipitation of $CaCO_3$. Calcium is involved in the formation of skeletons of certain algae and can be deposited in or on the cell walls of several algae as calcite or aragonite. To a certain extent, calcium can be replaced by strontium. However, this replacement may lead to differences in cell-wall composition or to an increase in cell size, so it appears that some functions of calcium can be carried out by strontium whereas others seem not to be mediated by this element.

Sodium and potassium

Sodium appears to be essential for cyanobacteria only, and is required by some algae and not by others. It has been suggested that higher concentrations of sodium might be toxic to cyanobacteria, which may account for the lack of these algae in marine environments. In contrast to this hypothesis is the abundant growth of *Spirulina* and other cyanobacteria in inland saline lakes, and it has to be assumed that Na is necessary for all marine and halophilic algae. It has also been proposed that sodium is required in nitrogen-fixing algae for the transformation of molecular nitrogen into ammonia.

Because sodium and potassium have similar chemical characteristics it has been assumed that sodium may replace potassium. Potassium is a requirement for all algae; under potassium-deficient conditions, growth and photosynthesis are reduced and respiration is high. Normality in such cells is restored by the addition of potassium or, to a certain extent, rubidium. This nutrient is a cofactor for several enzymes and is involved in protein synthesis and osmotic regulation. Providing sufficient potassium, the ratio of K^+ to Na^+ within the cell is independent of the ratio in the medium; algae that are deficient in K^+ tend to replace intracellular K^+ with Na^+.

Magnesium

Because of the strategic position magnesium occupies in the photosynthetic apparatus as the central atom of the chlorophyll molecule, all algal species have an absolute requirement for this element. Another key function of Mg is its role in the aggregation of ribosomes into functional units and for the formation of catalase. The minimum requirement of magnesium appears always to be greater than that for calcium. If both elements are provided in sufficient amounts, many algae tolerate a wide range of Ca/Mg ratios; this tolerance is connected with the tolerance to different ratios of monovalent/ divalent ions. In several chlorophyceae, magnesium deficiency interrupts cell division, resulting in abnormally large etiolated cells.

Iron

The importance of iron for the growth of algae is well substantiated: it is a key element in metabolism, being a constitutent of cytochromes; it plays an important role in nitrogen assimilation as a functional part of ferredoxin and affects the synthesis of phycocyanin and chlorophyll. Bleaching and yellow coloring of algae cultures are often an indication of iron deficiency in the medium.

The mode of supply of this element in the culture medium is still a subject of controversy, because it is not clear what fractions of particulate, colloidal or soluble iron are available to the algae. For a long time the supply of iron to algae cultures was critical because the iron, added in the form of inorganic salts, tended to precipitate, thus becoming unavailable to the algae. Nowadays, iron is mostly supplied in the form of chelated complexes, preferably bound to ethylenediaminetetraacetic acid (FeEDTA). These compounds are extremely stable, but release enough iron ions by mass action to satisfy the requirements of the algae. The use of iron as a citrate complex has also been reported, where the citric acid acts as a complexing as well as a reducing agent and as a pH buffer.

Trace elements

No medium can be considered adequate unless it is provided with several micronutrients. These elements are required in very small amounts of micro-, nano- or even picograms per liter. Trace elements are those elements that a) influences growth in a *representative number* of species, b) have a *positive* effect on total growth, c) show a *direct* physiological effect on alga growth, d) *cannot be replaced* by another element, and e) show *reversible* signs of deficiency in cultures lacking this element. The major trace elements in algal media are manganese, nickel, zinc, boron, vana-dium, cobalt, copper and molybdenium. Among the heavy-metal micro-nutrients there is often only a very narrow gap between a nutritional, growth-promoting effect and cell toxicity. For instance copper, essential at extremely low concentrations, is an algal poison at higher concentrations, and is even used in combating algal blooms. Similar conditions are also true for nickel. As already mentioned, molybdenium is an essential element for nitrogen-fixing cyanobacteria. Cobalt, bound in vitamin B_{12}, has been shown to be essential for a large number of species. While discussing the effects of trace metal ions on algal growth, further complex aspects have also to be considered, because variations in metal absorption and transport are often a function of the physiological state of the algal cell.

Several metal ions, present together in the culture medium, may affect the algal growth either by competitive mechanisms, by synergistic or antagonistic reactions, i.e. (toxic) effects, that are either additively intensified or mitigated so that it often becomes very difficult to predict the effect of a single metal ion in the culture medium without considering the influence of the other metals. Finally, the tremendous capacity of algal cells to accumulate heavy metals has to be mentioned in this context. Depending on the concentration in the medium, many algae are able, within short periods of time, to concentrate certain heavy metals more than 1000-fold within their cells compared with the outer concentration in the medium. This unique ability has already been considered as a means to detoxify heavy-metal-polluted waste water (see Chapter 14).

Many artificial culture media differ from what the algae would have in their natural environment because the concentration of most nutrients is far higher than the natural levels. These high concentrations are desirable for laboratory batch cultures where the biomass increases considerably without replacement of the medium in order to prevent growth limitation due to the deficiency of some mineral elements.

For large-scale cultivation, the concentration of the various elements has to be monitored more carefully in order to avoid precipitation due to changes of the culture conditions as well as for economic reasons, since the chemicals needed for the preparation of the media represent a considerable part of the production costs. If the culture medium has to be discarded, the loss would be greater if the concentration of the nutrients were unnecessarily high. Frequent estimation of the mineral composition and replenishment of the amount taken up by the algae is the method of choice for outdoor cultivation.

For the preparation of growth media for laboratory cultures, distilled water and analytical-grade chemicals are recommended, although special investigations may require additional methods for purifying the chemicals or the use of highly purified salts. An important prerequisite for the growth of axenic cultures is an effective method for sterilization of the culture medium. Two methods are commonly used for this purpose, namely heat sterilization and filtration. Heat sterilization, i.e. autoclaving under pressure, is performed usually for 20 min at 121 °C. This method cannot be used for the sterilization of solutions containing thermolabile substances such as vitamins, proteins or sugars. Autoclaving may also lead to the precipitation of certain mineral salts; for instance the tendency to flocculation in media rich in potassium salts, resulting in amorphous precipitates that render the

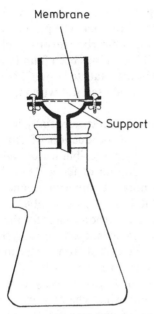

Membrane

Support

Fig. 3.7. Type of glass vessel with membrane filter for sterile filtration of culture media.

solution turbid. To overcome this problem the different salts have to be sterilized in different solutions that are subsequently mixed, or sterilized by filtration. The latter method is also recommended for the thermolabile solutions. Many types of filter are available for this purpose, the common feature of all of them being that the pore size is small enough to prevent the passage of bacteria. In practice cellulose membranes with pore sizes of 0.45 or 0.2 μm proved to be good. A standard assembly for membrane filtration is shown in Fig. 3.7, consisting of a glass filter mounted on a suction flask and heat sterilized. The medium to be sterilized is poured into the filter and a suction is applied. Other designs of such units also permit the application of pressure on the medium thus pushing the solution through the filter.

Unconventional sources of nutrients

Several algal projects, especially those located in developing countries, tend to modify and simplify the usual culture media in such a way that they suit the local conditions. Some of these media can then be applied at a more rural level, predominantly for the production of algae as animal feed and as a step to greater flexibility with the aim of simplifying the cultivation and processing procedures. These media permit the development of integrated algal production systems in which agricultural or other wastes could be utilized.

Some of these simplified media reported in the literature for the production of *Spirulina* are summarized in Table 3.4. The primary aim of these recipes was to reduce the high amount of bicarbonate provided in the standard medium described by Zarrouk (1966). It has been found that it is possible to reduce the amount of sodium bicarbonate to about 25% and to provide the trace elements in the form of crude sea salt without causing reduction in algal growth.

In the past decade there has been great interest, especially in several Asian countries, in using cow dung for the production of biogas. In this connection there also appear to be two possibilities of indirectly using dung for algal production: 1) the addition of fermented effluents, drained off from the biogas digester as nutrient supplement; and 2) the production of CO_2-enriched air ('aerobic bio-gas') by aerobically composting the dung and using the evolved gas as carbon source for the algae. For producing larger amounts of this gas in a continuous system, a prototype of a rural fermenter has been described from India that can hold about 1000 kg of dung and in which up to 140 g $CO_2 h^{-1} kg^{-1}$ dung can be produced by mixing fresh cow dung with actively fermenting compost. The gas coming out of this composting chamber can support growth of 5000 l of algal culture. Half of the composted dung has to be replaced with an equal amount of fresh dung every 8–10 days in order to keep up the CO_2 concentration. This fermenter consists of a cubic chamber made of bricks and mortar painted black inside for warming up the fermenting content, covered by a glass plate. The CO_2-enriched gas is drawn by a circulation device from the fermenter and bubbled into the algal culture.

Effluent from cow-dung biogas digesters as substrate has been used to replace certain chemicals in *Spirulina* culture media. The effluent was incorporated into a standard medium at 1–5% levels. At higher levels the initial growth rate of *Spirulina* was very low as the dark brown colour of the effluent reduced the amount of radiation into the medium and hence decreased the photosynthetic efficiency. Incorporating 3% of biogas effluent gave very satisfactory results.

Bone meal is a rich source of phosphate and calcium and is superior to mineral phosphates salts because it also contains certain amounts of nitrogen. Bone meal can be used as a substitute for calcium and phosphate salts in *Spirulina* cultures that are usually supplied with K_2HPO_4 and $CaCl_2$. It was found that the best method of providing this substrate is not by direct addition into the medium but by filling the required amount into a nylon sack immersed into the culture; 1–3 g l^{-1} were sufficient to maintain normal growth.

Another waste product that has been used in algal cultivation is urine. Cow urine (20 ml l^{-1}) was used successfully in outdoor cultures of *Spirulina* to replace nitrate.

It has been reported that small amounts of blood increase the growth of

Scenedesmus obliquus, cultivated in a simplified medium. Efforts were made to establish the quantitative effect of blood addition by identifying the active ingredients responsible for the effect and to explore its practical applications in large-scale cultivation. Whole sheep blood, plasma, hemolyzed cells, and crude preparations of carbonic anhydrase were tested. The highest growth-stimulating effect was obtained with whole blood. Since blood cells contain various active metabolites including heme compounds and metal ions etc., it is likely that these have a total impact in promoting the growth cumulatively. Iron, for instance, is an important constituent of blood cells, covalently bound to the heme molecule. The positive effects observed on algal growth may be partly caused by a better availability of iron, which is supplied through the blood and released at a slow rate into the medium at levels sufficient for the algal cells to utilize it more efficiently than inorganic iron provided through mineral nutrients.

Several algal culture media commonly used in laboratory and outdoor cultures are compiled in Tables 3.5–3.10. For the cultivation of chlorophyceae (mainly *Chlorella* sp. and *Scenedesmus* sp.) a selection of various media is summarized in Tables 3.11–3.14. These media are composed of cheap raw or waste materials, which may help to reduce the cost for different minerals otherwise required.

Sea water

Sea water is an abundant natural source of most of the major and minor elements required for algal mass cultivation. Because several algal production plants are located close to the sea, one would expect intensive use of this resource. However, except for the cultivation of *Dunaliella*, very few reports describe the utilization of sea water as part of algal culture media.

It is known that several fresh-water algal species grow in high concentrations of sea water if certain mineral nutrients normally absent, or present in low concentrations, are supplied. Growth tests performed with 13 species of green algae on enriched sea water (1 g $NaNO_3$, 0.11 g KH_2PO_4 and trace elements per liter of sea water) showed responses that varied from tolerance of full-strength sea-water medium (e.g. *Chlorella*) to strong inhibition even at low concentrations of sea water (e.g. *Chlamydomonas, Coelastrum*).

The utilization of media with high salinity for the cultivation of *Chlorella* was reported from Hong Kong. The domestic sewage effluent of this city is characterized by having a relatively high salinity (1.5%) because sea water is supplied for flushing systems. It was found that a strain of *Chlorella salina*, isolated locally from estuarine water, is able to grow in this effluent. Growth of this alga was found to be comparable to that of cells grown in a complete artificial medium.

Possibilities of utilizing sea water as a medium for outdoor cultivation of *Chlorella* sp. and *Scenedesmus* sp. were also reported from Thailand. Because bacteria and parasitic fungi are the main causes of low algal growth

Table 3.4. *Composition of different culture media ($g\,l^{-1}$ unless otherwise given) for Spirulina*

Substance											
NaHCO$_3$	16.80	4.50	9.00	9.00	12.00	3.00	—	3.00	19.20	2.10	4.00
K$_2$HPO$_4$	0.50	—	0.50	0.13	0.10	0.30	0.50	0.07	0.50	0.50	—
NaNO$_3$	2.50	—	1.50	1.25	2.50	—	0.25	—	—	2.50	3.00
NaCl	1.00	—	—	0.50	1.00	0.80	1.00	—	—	1.00	—
MgSO$_4$.7H$_2$O	0.20	0.20	0.20	0.10	0.09	—	0.15	—	—	0.20	0.40
FeSO$_4$.7H$_2$O	0.01	0.01	0.01	0.005	—	—	0.01	0.005	0.01	0.01	0.01
K$_2$SO$_4$	1.00	1.00	1.00	0.50	—	—	1.00	—	—	1.00	0.50
CaCl$_2$.2H$_2$O	0.04	—	0.04	0.02	0.25	—	0.04	—	—	0.04	—
EDTA	0.08	—	—	—	0.001	—	0.08	—	—	0.08	—
FeCl$_3$.6H$_2$O	—	—	—	—	0.80	—	—	—	—	—	—
Urea	—	—	—	—	—	0.30	—	—	0.20	—	—
Saltpeter	—	—	1.00	1.25	—	—	—	2.20	—	—	—
Sea salt	—	—	—	—	—	—	—	9.10	—	—	5.00
Cow urine	—	20 ml l^{-1}	—	—	—	—	—	—	—	—	—
Biogas effluent	—	50 ml l^{-1}	—	—	—	—	—	—	—	—	—
Bone meal	—	3.00	—	1.00	—	—	—	—	—	—	—
Cattle manure	—	—	—	—	—	40 ml l^{-1}	—	—	—	—	—
Sewage	—	—	—	—	—	75 ml l^{-1}	—	—	—	—	—
Swine blood	—	—	—	—	—	60 ml l^{-1}	—	—	—	—	—
HEPES buffer	—	—	—	—	—	—	20mM	—	—	—	—
A$_5$ solution[a]	1 ml l^{-1}	—	—	0.5 ml l^{-1}	—	—	1 ml l^{-1}	—	—	1 ml l^{-1}	—
B$_6$ solution[a]	1 ml l^{-1}	—	—	—	1 ml l^{-1}	—	—	—	—	1 ml l^{-1}	—
Water source	fresh	fresh	fresh	fresh	fresh	fresh	fresh	fresh	sea water	sea water	sewage
Aeration	no	no	no	no	no	air	air + CO$_2$	air	air + CO$_2$	no	no
Algal strain[b]	A	B	B	B	A	A	B	A	A	C	A

Notes:

[a] Composition of trace elements solution A_5 and B_6.

Substance	Amount ($g\,l^{-1}$)
A_5 solution	
H_3B_3	2.86
$MnCl_2.4H_2O$	1.80
$ZnSO_4.7H_2O$	0.22
MoO_3	0.01
$CuSO_4.5H_2O$	0.08
B_6 solution	($mg\,l^{-1}$)
NH_4VO_3	22.9
$NiSO_4.7H_2O$	
$Na_2\,WO_4$	18.0
$Ti_2(SO_4)_3$	40.0

[b] A, *Spurulina maxima*; B, *Spirulina platensis*; C, *Spirulina subsala*.

Table 3.5. *Medium for* Chlorella *and* Scenedesmus *sp*

KNO_3	810 mg l^{-1}
$NaNO_3$	680 mg l^{-1}
$NaH_2PO_4.H_2O$	415 mg l^{-1}
$MgSO_4.7H_2O$	250 mg l^{-1}
$Na_2HPO_4.2H_2O$	180 mg l^{-1}
$Ca(NO_3)_2.4H_2O$	25 mg l^{-1}
FeEDTA	4 mg l^{-1}
H_3BO_3	2.5 mg l^{-1}
$ZnSO_4.7H_2O$	1 mg l^{-1}
$MnCl_2.4H_2O$	0.2 mg l^{-1}
$CuSO_4.5H_2O$	0.1 mg l^{-1}
$(NH_4)_6Mo_7O_{24}.4H_2O$	0.03 mg l^{-1}

Table 3.6. *Culture medium for* Cyanidium caldarium

$(NH_4)_2SO_4$	1.32 g l^{-1}
KH_2PO_4	272 mg l^{-1}
$MgSO_4.7H_2O$	250 mg l^{-1}
$CaCl_2.2H_2O$	75 mg l^{-1}
FeEDTA	8 mg l^{-1}
$MnCl_2.4H_2O$	2 mg l^{-1}
H_3BO_3	3 mg l^{-1}
$ZnSO_4.7H_2O$	0.2 mg l^{-1}
$CuSO_4.5H_2O$	0.1 mg l^{-1}
$(NH_4)_6Mo_7O_{24}.4H_2O$	0.1 mg l^{-1}

Note:
pH 2.0 adjusted with HCl.

Table 3.7. *Medium for* Porphyridium cruentum

NaCl	27.0 g l^{-1}
$MgSO_4.7H_2O$	6.6 g l^{-1}
$MgCl_2.6H_2O$	5.6 g l^{-1}
$CaCl_2.2H_2O$	1.5 g l^{-1}
KNO_3	1.0 g l^{-1}
KH_2PO_4	70 mg l^{-1}
$NaHCO_3$	40 mg l^{-1}
FeEDTA	25 mg l^{-1}
H_3BO_3	0.6 mg l^{-1}
$MnCl_2.4H_2O$	0.4 mg l^{-1}
$(NH_4)_6Mo_7O_{24}.4H_2O$	0.4 mg l^{-1}
$CuCl_2.2H_2O$	0.04 mg l^{-1}
$ZnCl_2$	0.04 mg l^{-1}
1M Tris buffer pH 7.6	20 ml l^{-1}

Table 3.8. *Culture medium for* Dunaliella bardawil

NaCl	117.0 g l^{-1}
$NaHCO_3$	1.7 g l^{-1}
$MgSO_4.7H_2O$	1.2 g l^{-1}
KNO_3	505 mg l^{-1}
$MgCl_2.4H_2O$	100 mg l^{-1}
$CaCl_2$	33 mg l^{-1}
KH_2PO_4	14 mg l^{-1}
$ZnCl_2$	14 mg l^{-1}
H_3BO_3	6 mg l^{-1}
$CoCl_2.6H_2O$	4.8 mg l^{-1}
$CuCl_2.2H_2O$	$34 \text{ } \mu\text{g l}^{-1}$
Tris-HCl	6 g l^{-1}

Note:
Adjust the medium to pH 7.5 and then add 50 ml of iron solution (0.32 mg of $FeCl_3$ and 5.84 mg of Na_2EDTA dissolved in 50 ml distilled water).

Table 3.9. *Modified Allen's medium for cyanobacteria*

$NaNO_3$	1500 mg l^{-1}
K_2HPO_4	40 mg l^{-1}
$MgSO_4.7H_2O$	75 mg l^{-1}
Na_2CO_3	20 mg l^{-1}
$Ca(NO_3)_2.4H_2O$	20 mg l^{-1}
$Na_2SiO_3.9H_2O$	60 mg l^{-1}
EDTA	1 mg l^{-1}
Citric acid	6 mg l^{-1}
$FeCl_3$	2 mg l^{-1}
Microelements[a]	1 ml l^{-1}

Note:
[a] Composition of microelement stock solution:

H_3BO_3	2.90 g l^{-1}
$MnCl_2.4H_2O$	1.80 g l^{-1}
$ZnSO_4.7H_2O$	0.220 g l^{-1}
$Na_2MoO_4.2H_2O$	0.390 g l^{-1}
$CuSO_4.5H_2O$	0.080 g l^{-1}
$Co(NO_3)_2.6H_2O$	0.050 g l^{-1}

Table 3.10. *Allen and Arnon medium for cyanobacteria*

$MgSO_4.7H_2O$	240 mg l^{-1}
NaCl	230 mg l^{-1}
K_2HPO_4	350 mg l^{-1}
$NaNO_3$	210 mg l^{-1}
KNO_3	250 mg l^{-1}
$CaCl_2$	56 mg l^{-1}
Microelements[a]	1 ml l^{-1}

Note:
[a] Microelement stock solution:

FeEDTA	4000 mg l^{-1}
$MnSO_4.4H_2O$	500 mg l^{-1}
H_3BO_3	500 mg l^{-1}
MoO_3	100 mg l^{-1}
$ZnSO_4.4H_2O$	50 mg l^{-1}
$CuSO_4.5H_2O$	20 mg l^{-1}
NH_4NO_3	10 mg l^{-1}
$Co(NO_3)_2.6H_2O$	10 mg l^{-1}
$NiSO_4.6H_2O$	10 mg l^{-1}

Table 3.11. *Simplified culture medium for Dunaliella sp.*

Sea water	$1 \, l^{-1}$
Sea salt	$150 \, g \, l^{-1}$
Bird Dung (guano)	$20 \, mg \, l^{-1}$ (as PO_4^{3-})
Saltpeter	$450 \, mg \, l^{-1}$ (as NO_3^-)
Sodium bicarbonate	$500 \, mg \, l^{-1}$ (as HCO_3^-)
FeEDTA (10%) solution	$0.1 \, ml \, l^{-1}$

Table 3.12. *Simplified culture medium for Phaeodactylum sp.*

Sea water	$0.9 \, l^{-1}$
Fresh water	$0.1 \, l^{-1}$
Saltpeter	$50 \, mg \, l^{-1}$
$(NH_4)_2HPO_4$ (10%) solution	$0.1 \, ml \, l^{-1}$
$NaHCO_3$	$40 \, mg \, l^{-1}$
Mine salt	$200 \, mg \, l^{-1}$
FeEDTA (10%) solution	$0.1 \, ml \, l^{-1}$

Table 3.13. *Simplified culture media for chlorophyceae ($mg \, l^{-1}$ unless otherwise given)*

Urea	100	—	—	600	300
K_2CO_3	25	—	—	—	—
Na_3PO_4	25	—	—	—	—
$CaCl_2.2H_2O$	—	—	—	—	7
$(NH_4)_2SO_4$	20	—	—	—	—
$MgSO_4.7H_2O$	10	—	—	300	120
$FeSO_4.7H_2O$	2	—	—	5	10
K_2HPO_4	—	250	50	—	144
KH_2PO_4	—	—	—	300	92
EDTA	—	—	—	7	—
Crude salt	25	—	—	—	—
Acetic acid	—	—	—	$3–6 \, ml \, -^{-1}$	—
Saltpeter [a]	—	750	500	—	—
Yeast waste[b]	—	—	$30 \, ml \, l^{-1}$	$30 \, ml \, l^{-1}$	—
A_5 solution[c]	—	—	—	—	$1 \, ml \, l^{-1}$
B_6 solution[c]	—	—	—	—	$1 \, ml \, l^{-1}$
Water	fresh	fresh	sea	fresh	fresh

Notes:
[a] Saltpeter: $NaNO_3$, 75%; KNO_3, 21%; NaCl, 1%; $NaSO_4$, 1.25%.
[b] Yeast waste: total solids, 4.5%; volatile solids, 3.4%; $NH_4–N$, 0.07%; PO_4^{2-}, 0.3%.
[c] A_5 and B_6 composition: see Table 3.4.

Table 3.14. *Simplified inorganic medium for cyanobacteria and chlorophyceae*

Demineralized water	890 ml
Sea water (35–38‰ salinity)	100 ml
KNO_3	300 mg l^{-1}
$K_2HPO_4.3H_2O$	25 mg l^{-1}
Trace element solution A	5 ml l^{-1}
Trace element solution B	5 ml l^{-1}
pH: *c.* 7.5	
Trace element solution A:	
$MnCl_2.4H_2O$	20 mg l^{-1}
$ZnSO_4.7H_2O$	5 mg l^{-1}
$CoSO_4.7H_2O$	5 mg l^{-1}
$Na_2MoO_4.2H_2O$	5 mg l^{-1}
$CuSO_45H_2O$	0.5 mg l^{-1}
Trace element solution B:	
$FeCl_3.6H_2O$	1 mg l^{-1}
$Na_2EDTA.2H_2O$	275 mg l^{-1}

Source: After Pohl *et al.*, 1987.

rates or complete breakdown of the cultures in this area, an attempt was also made to find out the applicability of using sea water for preventing and controlling these infections. For the cultivation of a *Chlorella* strain isolated locally, a medium composed of 20 g sea salt per liter plus urea, $FeSO_4.7H_2O$ and plant fertilizer (N:P:K:Mg = 12:12:17:2) at a concentration of 0.5 mM P, 4.3 mM N, and 0.01 mM Fe respectively, gave results that exceeded even the results obtained with fresh-water media. Similar experiments conducted with *Scenedesmus obliquus* (strain 276–3a) revealed that with this alga productivity and protein content decreased with increasing salt concentrations. In addition, at salinity above 15‰ the algal cells became pale and the volume increased so that the original characteristics of the alga were transformed beyond recognition. A new and inexpensive inorganic medium for the mass cultivation of fresh-water cyanobacteria and chlorophyceae, based on sea water, has been published recently (Pohl *et al.*, 1987). It consists of tap water (90%) with a low calcium content (40–80 mg CaO l^{-1}), sea water (10%), K_2HPO_4 (25 mg l^{-1}), KNO_3 (300 mg l^{-1}) and 20 ml l^{-1} of each trace element solution, the compositions of which are given in Table 3.15. The medium has a low phosphate and trace-element concentration and does not require the addition of boron because this element is provided in sufficient quantities by the sea water. According to the authors, the medium leads to rapid algal growth and has been tested with numerous chlorophyceae and cyanobacteria grown in bioreactors.

Another approach for testing the feasibility of sea water or other cheap

Table 3.15. *Artificial sea water for*
Porphyridium sp.

NaCl	27.0 g l^{-1}
$MgSO_4.7H_2O$	6.6 g l^{-1}
$MgCl_2.6H_2O$	5.6 g l^{-1}
$CaCl_2.2H_2O$	1.5 g l^{-1}
KNO_3	1.0 g l^{-1}
KH_2PO_4	70.0 mg l^{-1}
$NaHCO_3$	40.0 mg l^{-1}
Tris CHl (1 μmol, pH 7.6)	20 ml l^{-1}
Fe + EDTA (10%) solution	1 ml l^{-1}
Microelements	1 ml l^{-1}

Note:
Composition of microelement stock solution:

$ZnCl_2$	4.0 mg l^{-1}
H_3BO_3	6.0 mg l^{-1}
$MnCl_2.4H_2O$	40.0 mg l^{-1}

raw materials as components of algal culture media is to prepare a simple medium in the first instance and then to test a series of species and strains of potential algae for their growth performance in order to select the best form.

Detailed work on the biochemical taxonomy of *Chlorella* has revealed the existence of a wide variety of biochemical properties and physiological tolerances within this seemingly uniform, but actually heterogenous, genus of chlorophyceae. Thirty one *Chlorella* strains, belonging to seven species, were tested in a medium based on sea water enriched with nitrate and phosphate (KNO_3, 810 mg l^{-1}; $NaHPO_4.12H_2O$, 360 mg l^{-1}; $NaH_2PO_4.2H_2O$, 470 mg l^{-1}). The productivity was compared with that obtained in the normal culture medium used for laboratory cultures. Strains of only two species seem to be halophilic rather than halotolerant, i.e. they grow better in sea water than in the normal medium. Although strains of other species are less productive in sea water, some of them are able to produce acceptable yields. Those strains that gave at least 50% of the yield obtained in the control medium are listed in Table 3.16. The results indicate that by screening the different strains available nowadays from various culture collections it might be possible to select strains that have the capacity of producing high yields in simple media, of which sea water (or sea salt) is probably the cheapest component.

Phosphate estimation

For proper algal pond management, routine estimates of phosphate concentration in the algal medium are necessary. In the following list, a simple photometric method for phosphate estimation is given.

Table 3.16. *Growth of different strains of six* Chlorella *species in normal medium and enriched sea water medium; production of dry matter after cultivation for eight days (25°C, 8000 lux)*

| Species | Strain[a] | Yield (g l^{-1}) | | % Growth in sea water |
		Normal medium	Sea water	
C. sorokiniana	211–31	0.80	0.53	66
C. vulgaris	211–81	0.67	0.58	87
	211–8m	0.45	0.33	73
	211–11q	0.51	0.40	78
	211–11s	0.52	0.38	73
	211–12	0.49	0.28	57
	395	0.33	0.33	100
C. saccharophila	211–9a	0.59	0.50	85
	211–9b	0.09	0.23	256
	211–1b	0.46	0.71	154
	211–1c	0.59	0.69	117
	211–1d	0.63	0.68	108
	211–1f	0.16	0.41	256
C. luteoviridis	211–2a	0.07	0.12	172
	211–2b	0.40	0.22	55
	211–3	0.40	0.26	65
	211–4	0.31	0.42	136
	211–5a	0.20	0.27	135
	211–5b	0.26	0.32	123

Note:
[a] Strain numbers according to List of Cultures, Algal Culture Collection, University of Göttingen (Germany).

The reagents used are as follows.

a) 95% sulfuric acid;
b) 17.3% sulfuric acid;
c) molybdate–vanadate solution (solution A).
   Preparation of solution A:
   Add 12.5 ml of 95% H_2SO_4 to 40 ml distilled water, cool in ice. Weight 2.5 g $(NH_4)_6Mo_7O_{24}.4H_2O$ and 0.1 g of NH_4VO_3 into 40 ml distilled water. Dissolve the salts by heating to 50–60 °C, after that cool the solution. Add the H_2SO_4 solution to the salt solution and make up the volume to 100 ml with distilled water. Store the solution in a brown bottle in a cool place. Procedure: to a 5 ml filtered sample add 0.2 ml of 17.3% H_2SO_4 and add 0.4 ml of solution A. Mix well and allow the mixture to stand for 10 min before reading the absorbance at 365 nm against a reagent blank.

A modification of the above methods is the following.

1) 10 g ascorbic acid are dissolved in 100 ml distilled water (store at 2–4 °C).
2) Dissolve 2.5 g $(NH_4)_6Mo_7O_{24}.4H_2O$ in 100 ml distilled water.
3) Prepare 6 N H_2SO_4 by carefully adding 18 ml of concentrated H_2SO_4 to 90 ml distilled water.
4) Prepare reagent A by mixing 10 ml 6 N H_2SO_4, 20 ml distilled water, 10 ml molybdate solution and 10 ml ascorbic acid solution. This reagent has to be prepared fresh daily since it is not stable.
5) Mix 1 ml of sample and 4 ml of solution A together in a test tube. The closed tube is kept for 2 h at 37 °C. After cooling at room temperature the absorbance of the formed complex is determined at 820 nm.

The read absorbance is compared with a standard graph prepared from a known amount of phosphate (i.e. 143 mg $KH_2PO_4\,l^{-1} = 100$ mg $PO_4^{2-}\,l^{-1}$).

4   Cultivation methods (indoors)

Various methods have been developed to isolate and cultivate microalgae in the laboratory as well as outdoors on a large scale. In the following, a brief survey is given of the basic laboratory techniques and equipment necessary for the cultivation of algae.

For continuous cultivation, isolation steps or multiplication procedures, the algae have to be transferred from one medium to another. The commonly used appliances for this are wire needles made out of platinum or chromium with the tip twisted to form a small loop, glass rods, capillary tubes, or small blades, mainly for cutting out agar and for severing mats of filamentous algae. The streak method, commonly used in microbiology for the isolation of bacteria, can also be adapted for the isolation of algae. For such cultures 20 ml of 2% washed agar in a nutrient medium are poured into sterile Petri dishes and allowed to solidify. A dumb-bell tipped bent glass rod is dipped into the algae suspension and streaked on the surface of the agar at two or three places (Fig. 4.1). By this method, single cells or groups of cells are left behind as the rod moves, so that discrete colonies may develop.

Culture vessels

Usually culture vessel made out of glass are preferred except in particular cases where other special types are required. For instance, glass vessels are not suitable for studies on silicon nutrition, because glass contains various silicates that are all soluble to a certain extent. For such purposes vessels made from polythene or other modern plastics may be used. The culture vessels are usually stoppered by means of cotton plugs wrapped in cotton gauze and covered by wax paper or aluminum foil, by rubber stoppers or by aluminum caps. The latter are commonly used for stock cultures kept on agar slants in test tubes.

For routine cultivation of small volumes of algal culture, test tubes will suffice. If larger quantities of algae are required or if the ratio of volume to surface has to be increased, 200–300 ml Erlenmeyer flasks or flat-bottomed depressed flasks (Fig. 4.2A) are usually employed, filled by approximately 30% of their volume with the algal suspension. The flasks are closed with a cellulose or cotton-wool plug and preferably covered with aluminum foil to exclude dust and to reduce evaporation. The advantage of this method lies

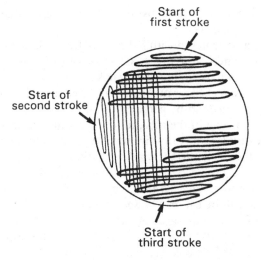

Fig. 4.1. Method of streaking on agar plates.

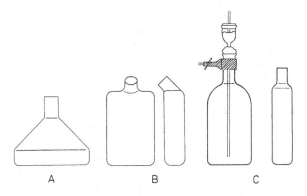

Fig. 4.2. Different types of glass vessel for laboratory culture of algae. A, Flat-bottomed depressed vessel; B, Penicillium flask; C, 'Roux'-flask, with and without sealed-in aeration tube.

in the ease with which axenic cultures can be kept over long periods of time. Filamentous forms, however, occasionally form clumps which make the quantitative withdrawal of samples difficult.

These conical flasks are suitable for cultures of up to 250 ml, which can be kept in rotary shakers, and also installed in illuminated, temperature controlled and aerated incubation chambers. The 'Penicillium flask' (Fig. 4.2B) is suitable for growing cultures up to 1 l under ordinary laboratory conditions. The main disadvantages of the cultivation in Erlenmeyer or similar flasks are the unfavourable illumination due to shading by stoppers and supports, and the limited passive gas exchange through the plugs which results in deficient CO_2 supply in dense cultures. A direct gas supply into the cultures by means of pipes and tubes is possible but makes the system too clumsy if it is installed on shakers.

Higher irradiance can be achieved with flat flasks ('Roux' flasks) which offer a larger surface for illumination (Fig. 4.2C). However, as the geometry of this flask does not allow it to be mounted in a shaker, the cultures have to be agitated by other means. A glass air tube can be inserted to the bottom of the flask for agitating the medium by aeration. A short length of plastic tubing is fitted over the top end of the tube, followed by an adjustable air valve and a long plastic tube leading to an air pump. A cotton plug is wound around the air tube and wedged into the opening of the bottle.

The advantage of this design, compared with round flasks, is an increased ratio of surface to volume, so that a more uniform and intense illumination of the algal culture can be obtained. In order to achieve high growth rates, mixing of the culture is essential for most of the algae. It prevents the algae cells from settling on the bottom of the culture vessel and ensures a uniform exposure of the individual cells to light and nutrients. In many culture devices, mixing of the medium is performed by aeration, where a stream of filtered sterile air or an air/CO_2 mixture is continuously bubbled through the culture.

For mixing small volumes kept in cylindrical flasks, the most common device is the magnetic stirrer, using small Teflon-coated magnetic bars. When several flasks or bottles have to be agitated simultaneously, rotary shakers can be used kept under a board with fluorescence tubes or similar installations. If continuous aeration is necessary, flasks with sealed-in aeration tubes are very useful. A series of such cultivation tubes can be incubated in a thermostated water bath with illumination provided from the sides.

Fig. 4.3 shows a culture tube that has been used successfully in several laboratories. It combines some advantages, i.e. reduction of time required to prepare the tube for use, ease of maintenance and economical construction. The tube has an inner diameter of 3.5 cm and a length of 40 cm, holding about 250–300 ml of algal suspension. Through a glass pipe attached to the bottom of the tube, CO_2-enriched air is bubbled through the culture to supply carbon and to prevent settling of the algae. The tube is closed with a porous cellulose stopper covered by an aluminum cap. The gas pipe introduced into the bottom is 4 cm shorter than the tube to enable easy flaming of the edge of the tube after rebending the plastic pipe connected to the cotton-filled bulb, which serves as a sterile filter for the incoming air.

The tubes of algal cultures are usually inserted into a thermostatically controlled water bath (Fig. 4.4) and a panel of fluorescent tubes is used for the illumination, the degree of which can be varied by changing the number of tubes and/or their distance from the culture. Heavy foam formation by some algal species can be encountered by the addition of anti-foam agents.

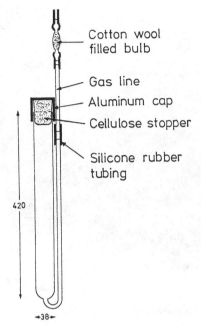

Cotton wool
filled bulb

Gas line

Aluminum cap

Cellulose stopper

Silicone rubber
tubing

420

◄38►

Fig. 4.3. Glass tube for indoor algal cultivation (dimensions in mm).

Fig. 4.4. Light and thermostatically controlled apparatus for indoor algal
cultivation in aerated glass tubes.

Fig. 4.5. *Spirulina* cultures raised in carboys.

When scaling up a new algal culture it is often desirable to keep the starter inoculum for outdoor ponds in shade rather than exposing it directly to sun light. This has been found very useful especially for the cultivation of *Spirulina*. When the cultures in the smaller glass tubes start growing after a few days, as evidenced by thickening of the culture and development of an intense blue-green colour, it can be diluted into larger glass carboys (capacity 5–10 l) aerated by compressed air through sintered stones and shaken a few times every day (Fig. 4.5).

Since many algae tend to adhere to the glass surfaces and may even form a crust, which cannot be removed by simple rinsing or mechanical cleaning with brushes, a more thorough cleaning of the cultivation vessels is required. The ordinary practice is to soak the glassware for several hours in chromic acid or in a mixture of concentrated sulfuric acid and potassium dichromate, after which they are repeatedly rinsed with tap water and finally with distilled water.

5 Scaling up

Preliminary studies as part of the screening for suitable and profitable algal strains, which are worth being produced on a large scale, can be performed with small volumes (up to 10 l) of laboratory cultures, i.e. analyses of chemical composition and screening for valuable ingredients, optimization of medium composition and culture conditions, establishing product analysis methods, development of suitable harvesting, processing and/or extraction procedures. These initial studies will indicate whether or not further scaling up is sensible, bearing in mind, for instance, too specialized cultivation or processing steps.

If the results of the preliminary investigations are promising, it is advisable to proceed to the next step of scaling up by growing the alga in small outdoor units (let's say up to 500 l) such as experimental ponds, tubular reactors or simple plastic bags (as will be described later).

At this stage data on the effect of environmental factors (light, temperature and contamination) can be collected and the performance of agitation, aeration and culture medium composition studied. Even in established large-scale cultivation, small units of 500–1000 l should always be maintained to serve as fresh inoculum for the routine preparation of new cultures in the large production units. Then a decision has to be taken whether to proceed to the next step, which is to the final pond size envisaged, or whether another intermediate pilot-plant stage should be incorporated. The decision depends on several economical and technical factors.

6   Algal grouping

Formerly, photosynthetic organisms were grouped into three more-or-less distinct categories: photosynthetic bacteria, algae and higher plants. Photosynthetic bacteria perform anoxygenic photosynthesis, i.e. their light-dependent CO_2 assimilation is not accompanied by oxygen production. Instead they use the reduced materials hydrogen sulfide, thiosulfate or organic compounds as source for the required reducing power to reduce CO_2 into cellular material. Their light-absorbing bacterio-chlorophyll is chemically distinct from chlorophyll a, present in the other photosynthetic organisms.

The prokaryotic blue-green algae hold a position between this photosynthetic bacteria and the eukariotic algae. They do not possess chlorophyll b, the role of which is filled in by different phycobiliproteins. In contrast to the bacteria, the blue-green algae perform oxygenic photosynthesis. Based on these characteristics and the fact that their cell-wall constituents, ribosome structure and nucleic-acid structure are more similar to bacteria than to eukaryotic algae, they are nowadays called cyanobacteria. Hence, in the following the expression cyanobacterium will be used as far as possible for the organism formerly called a blue-green alga.

7   Strain selection

A wide variety of algal strains and species has been screened and tested at various places for their suitability to be grown in mass cultures. One of the selected strains is *Spirulina*, which offers promising potentials for large-scale production. Because a number of *Spirulina* species with differences in growth physiology, morphology and cell composition are being cultivated worldwide, improved classification and selection methods seem to be necessary in order to find out the optimum species. For several years there have been projects going on in Israel aimed at developing an optimal biotechnology for the mass production of *Spirulina*. In this context, different methods were tested for strain selection of *Spirulina*, but which due to their exemplary nature also could be applied to other algae as well (Vonshak, 1987).

In order to develop a more specific classification, besides obvious morphological differences, two chemical procedures were applied: comparison of protein pattern of the photosynthetic membrane by polyacrylamide gradient, and analysis of the fatty-acid content. However, because these parameter are significantly influenced by several environmental parameters, additional physiological criteria have been determined that are considered to be crucial for the evaluation of high production rates in outdoor cultures.

1) Response to diurnal fluctuations. In those areas that are suitable for large-scale algal cultivation because of their high light intensity, diurnal temperature fluctuations of up to 20 °C are frequent, which means that especially during the morning hours, despite sufficient light, the temperature in the cultivation pond is far below the optimum temperature for growth. To minimize this growth-depressing effect, algal strains with a wide optimum temperature range are advantageous.

2) Resistance to photoinhibition. Although in general light is one of the limiting factors in algal cultivation, the very upper layer of the culture might be exposed to such high solar radiation that photo-oxidative bleaching of the algal cells occurs. Hence the aim of the selection should be to isolate strains with high light-saturation values and minimal growth inhibition at high light intensities.

3) Amount of dark respiration. Measurements of dark respiration rates have shown that during the night up to 35% of the total

biomass produced during the day is lost. Algal strains with a low dark-respiration rate or a high ratio of light-dependent O_2 evolution to O_2 uptake in the dark should make good candidates for outdoor cultivation.

4) Sensitivity to high O_2 concentrations. Oxygen is one of the factors in the overall photosynthesis equation, with the effect that high O_2 concentrations suppress the rate of photosynthetic carbon incorporation in form of a feed-back reaction, sometimes even leading to the total loss of the culture. Besides ensuring homogeneous distribution of algal cells in the medium, one of the purposes of inducing turbulence in algal cultures is to remove from the culture photosynthetically produced oxygen, the concentration of which otherwise may reach levels 500% of saturation. As a result, strains that can tolerate high oxygen concentrations may be selected.

5) Sensitivity to osmotic stress. High evaporation rates (up to $10 \, l \, d^{-1} \, m^{-2}$) in outdoor cultures and the addition of fresh medium may result in a twofold increase of the salt concentration after extended growth periods, which may cause severe osmotic stress. Thus, algal strains that tolerate increasing osmotic pressure without increasing respiration rates should be preferred.

8　Growth kinetics

Strictly spoken, the word 'growth' in common usage is not the precise term to describe the increase of algal biomass. An organism is said to 'grow' when the size of its amorphous mass increases by the addition of compounds from the surroundings. Growth of a living organism is defined as an increase in mass or size accompanied by the synthesis of macromolecules, leading to the production of a new organized structure. A living tissue grows by an increase in the size of its component cells, which may be accompanied by an increase in their number by cell division and/or by the inclusion of organized interstitial cellular products. Populations of organisms, as is the case with unicellular algae, may also be said to grow because increase in number of organisms by replication is accompanied by synthesis and a new organized structure is produced. In multicellular organisms such as filamentous algae, growth may include differentiation to produce cells or organs of particular form or function. Several filametous algae, however, are little more than chains of individual organisms. Any cell of such a filament is potentially capable of binary fission and hence the growth in length of the chain is merely the sum of the growth of the individual cells.

Thus growth, whether of a single macromolecule in a living cell or of a complex plant, produces order out of chaos, materials being incorporated in a specific system under the control of the genetic material of the particular organism.

Generally, populations of unicellular organisms may be measured in terms either of the number of individual cells or of their mass. The former may be termed 'cell concentration', defined as the number of individual cells per unit volume, whereas the latter may be called 'cell mass' or 'cell density', defined as the weight of cells or biomass per unit volume.

Under the typical regime of a simple homogenous batch culture (closed system), where the food supply is limited and nothing is added or removed from outside, the algal growth passes through several different phases.

1) Adaption (lag phase);
2) accelerating growth phase;
3) exponential growth (log phase);
4) decreasing log growth (linear growth);
5) stationary phase;
6) accelerated death;
7) log death.

51

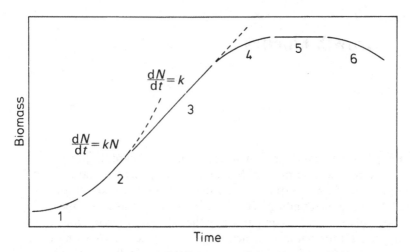

Fig. 8.1. Growth phases of algal cultures (details are given in the text).

The ideal shape of the respective growth phases is shown in Fig. 8.1. The individual phases are not always clearly pronounced; the different parts of the growth curve may be altered in length or slope according to the conditions prevailing in the culture. The various phases represent the reaction of the algal population to the changes of the environmental conditions and depend on the inoculum, the actual cultivation method, nutrient concentration, light intensity, temperature, etc.

Phase 1. When a medium is inoculated with organisms, the conditions at first are usually different from those of the previous environment from which they were taken. Often, the organisms are not adapted to the new environment and may even be in an unhealthy condition. During the adaption or lag phase, the algal culture adjusts itself to the altered conditions, and the specific growth rate is significantly lower than that in the subsequent one and increases with cultivation time. The length of the lag period is defined in different ways; the lag in cell counts, for instance, is greater than the lag in metabolic rate of reactions involved in growth.

Furthermore, the physiological age of the cells affects their capability for multiplication; that is, cells taken from late lag phase, log phase or early stationary phase will produce a shorter lag time than cells from the late stationary phase. High physiological activity is found during the lag phase, the cells being much more sensitive to temperature or other environmental changes than cells in a more mature stage of development.

Phases 2 and 3. After a short period of preparation or 'tooling up', where the algal culture has adapted itself to the given cultivation condition, the cells enter the phase of expentional (logarithmic) growth. During this period the light intensity is not limiting and changes in the nutrient concentration caused by the uptake by the algae is so small that this effect on the growth can be neglected. In such a culture, which is neither light nor nutrient limited, the increment in algal biomass (expressed as dry weight, cell number, optical density, etc.) per time is proportional to the biomass in the population at any given moment. A steady-state continuum is observed and the plot of the logarithm of cell mass yields a linear increase with time. The cells are dividing at a constant rate, determined both by the intrinsic nature of the organism and the culture conditions. Knowledge of the growth rate is important in determining the state of the culture. If one assumes the doubling of the initial biomass M_i in time g, the final concentration of the biomass M_f is:

$$M_f = M_i 2^n \tag{8.1}$$

where n is the number of cell divisions in time t.

The equation

$$g = t/n \tag{8.2}$$

gives the doubling time or mean generation time. This factor is the reciprocal of the growth rate constant, expressed as the number of doublings per unit time. The doubling time represents the average generation time of the culture as a whole, usually determined by doubling of the algal mass in the culture.

If we take the total growth time t_t measured from the time of inoculation, we have to consider the duration of the lag phase t_l plus the number of increments of the doubling time t_d that have occurred. Thus the corresponding times for each stage of the population growth will be:

$$t_t = t_l + n t_d. \tag{8.3}$$

This equation may be re-arranged into:

$$n = (t_t - t_l)/t_d. \tag{8.4}$$

Therefore, from equations (8.1) and (8.3) the following equation can be obtained:

$$M_f = M_i 2(t_t - t_l)/t_d. \tag{8.5}$$

This expression can be simplified by taking logarithms to the base 2 of each side of the equation, when:

$$\log_2 M_f = [(t_t - t_l)/t_d] + \log_2 M_i. \tag{8.6}$$

A plot of $\log_2 M_f$ against time gives a straight line from the point where the population is equal to the inoculum size at the end of the lag phase.

The doubling time g is best determined by calculation. To accomplish this, the increase of cell mass is determined in a known time interval and the generation time is calculated from the values obtained. Re-arranging equation (8.2) to the form

$$n = t/g \tag{8.7}$$

which is substituted into equation (8.1), we obtain:

$$M_f = M_i 2^{t/g}. \tag{8.8}$$

By conversion to the logarithmic form and re-arranging:

$$g = 0.69t/(\ln M_f - \ln M_i). \tag{8.9}$$

This equation expresses the doubling time from two measurements that give the increase of mass in time t. For calculating the specific growth rate the logarithmic form of equation (8.8) can be applied:

$$\ln M_f = t(\ln 2/g) + \ln M_i. \tag{8.10}$$

If we set $\mu = \ln 2/g$ we obtain:

$$\ln M_f = t\mu + \ln M_i, \tag{8.11}$$

where the constant μ is the slope of the straight line, when the values of t and $\ln M_f$ are plotted. This constant is called the 'specific growth rate' or 'instantaneous growth rate'. Its value can be determined as:

$$\mu = \ln 2/g = 0.69/g. \tag{8.12}$$

This factor μ is specific for every organism and culture medium and is governed primarily by the growth capacity of the organism and the environment.

Phase 4. The algal growth in a dense batch culture, where nothing is added to or removed from the medium after inoculation, proceeds through another phase where decreased logarithmic growth starts to occur. In this phase the cells have multiplied to such an extent that they begin to shade one another so that gradually a high absorption of incident light occurs. This effect, as well as the beginning of limitation of minerals and increasing accumulation of toxic wastes, reduces the specific growth rate and the increase in the algal biomass becomes almost linear. This phase continues until limitation due to exhaustion of a certain nutrient occurs or the culture reaches a stage where respiration begins to interfere. In nutrient-rich, well-maintained cultures, this linear phase remains over a certain period of time.

Phase 5. During the next phase, the light supply per algal cell becomes limited and respiration plays an increasing role. Oxidative breakdown of synthesized substances begin to reduce the constant increment of biomass and an equilibrium is reached between the maximum concentration of algal biomass and loss due to degradation processes. This phase is not distinct from the previous one but reached as a slowly increasing stage of the culture. The growth curve asymptotically approaches a limiting value, the stationary phase, which may be described as the maximum attainable concentration of algal biomass in a closed system.

Phase 6. The end of the stationary phase is marked by reduced viability in the cell population; the algal cells begin to die releasing organic, often growth-inhibiting, materials into the medium. This phase is caused by unfavourable environmental conditions, over-age of the culture and limited supply of light and nutrients or infection by other microorganisms.

Phase 7. Finally the death rate becomes exponential, leading to the complete breakdown of the algal population.

A special form of a culture in a closed system is the synchronized batch culture. In this type the growth is not linear but proceeds stepwise because all the cells divide almost simultaneously. This synchrony can be achieved by repeated shifts in environmental conditions (light, temperature, nutrient supply) or separation of cells from a random population and subsequent recultivation of all those cells that are at the same stage of age within the cell cycle. As mentioned above, these types of growth pattern occur in closed systems. In algal mass cultivation, however, continuous or semi-continuous cultivation (open system) is the desirable method for maintaining the culture in a steady-state (exponential) growth phase.

In the open system, fresh nutrients and all factors promoting algal growth are added continuously in the same amount that medium is withdrawn from the culture, in order to keep all growth paramater and cell concentration at a constant level. In such a system, the most important factor controlling growth of the organism is the rate at which fresh medium is added. This ratio of the flow rate f, at which fresh medium is added to the culture volume V is referred to as the dilution rate D, so that $D = f/V$.

This rate is the number of volumes of medium that pass through the cultivation pond in 1 h; its reciprocal, $1/D$, is the mean residence time of an algal cell in the system. The net change in the number of organisms with time is determined by the relative growth rate and the removal (wash out) from the pond:

$$dN/dt = \mu N - (f/V) \tag{8.13}$$

thus

$$dN/dt = \mu N - DN, \tag{8.14}$$

where μ is the instantaneous growth rate constant and N is the number of algal cells per volume.

For the size of the algal population to remain constant in the system, the following equations have to be fulfilled:

$$dN/dt = 0 \text{ and } \mu = D. \tag{8.15}$$

With a constant dilution rate all components are constant and a steady state is achieved.

9 Measurement of algal growth

One of the basic parameters for monitoring the performance of algal production systems is the estimation of the algal biomass produced. The growth of algal cultures is expressed usually as the increment of biomass, number of cells, amount of protein, pigments etc., over a given period of time. For this purpose various methods have been described in the literature and some of the procedures commonly used in algal cultivation will be given here (Vonshak, 1986).

Turbidity

Measurement of the turbidity is the most general method for estimating the algal concentration of cultures. Its advantage is the fast and non-destructive procedure. Two different techniques can be used:

1) a field method by visual observation;
2) a laboratory method requiring a colorimeter or spectrophotometer.

Method 1. A rough estimation of the density of the algal suspension, representing the concentration of suspended solids in the culture, can be obtained by the use of a so-called Secchi disk. This disk is a white-painted circular plate with a diameter of 10 cm, fixed to a calibrated stick. The disk is lowered from the surface into the algal culture until its white colour vanishes; the corresponding depth can be read from the stick. Because the visibility of the disk depends on the concentration of solids in the culture, a correlation can be established between this concentration and the respective depth. The measured extinction depth D_s is about one-half of the actual depth of light penetration D_p, because the light must enter the medium, strike the disk and then exit to be observed. According to Oswald (1988), in outdoor cultures D_s and D_p are related to the culture concentration approximately as: $C_c = 3000/D_s$ or $C_c = 6000/D_p$, in which C_c is the concentration of algae (mg/l^{-1}), D_s the Secchi depth (cm) and D_d the depth of light penetration (cm).

Field observations further indicate that the culture concentration in light-limited, well-mixed cultures approaches that which permits light to penetrate two-thirds of the actual culture depth, i.e. $D_p = (2/3)d$ and $C_c = 1.5 \times 6000/d - 9000/d$. A continuously mixed culture of depth 30 cm will obtain an average maximum light-limited algal concentration of about $9000/30 = 300$ mg/l^{-1}. Because the penetration of light is proportional to the logarithm of its intensity, the algal concentration would be only slightly higher during periods of high light intensity. This means that high algal concentrations can be obtained using only shallow culture depths.

The above method of estimating the algal concentration cannot be applied when the water itself is colored by water-soluble compounds, decomposed organic matter, or other dyes.

Method 2. A more elaborate method for estimating the density or concentration of suspended solids in a sample is measurement of the optical density at a given wavelength. The method is based on the application of the photometric law, stating that each elementary layer of a dispersed system scatters the same relative portion of the passing monochromatic light, the magnitude of the scatter being proportional to the dispersion concentration. Because the photometric law holds reliably also for suspensions of such particles as microscopic algae, it offers a rapid, simple and apparently accurate method for determining the concentration of cells in a suspension. However, this method can lead to erroneous results as the value of the scattering coefficient, indispensable for the correct use of the method, is a complex function of cell size, shape and content of scattering substances, and depends on the statistical distribution of the algal cells in the solution. These measurements may be carried out safely outside the region of the absorption of photosynthetic pigments. For this purpose, a wavelength around 550 nm is recommended. By plotting a standard graph correlating the light absorption and the dry weight at different concentrations, the amount of algal biomass in the sample can be estimated by simply measuring its absorption.

Dry-weight estimation

The gravimetric estimation of algal dry weight (sometimes called total suspended solids (TSS)), is one of the most direct ways to determine biomass production. However, this parameter only characterizes the general physico-chemical state of the given sample. The method cannot always be applied as the sole method for biomass estimation because it does not differentiate between the actual algal biomass and suspended non-biological solids (e.g. silt), non-algal planktonic material or minerals dissolved in the water because it includes all particles, which either remain on the filter used for the separation of the algae from the medium or which are collected by centrifugation of the algal sample. On the other hand, filtration does not necessarily retain all solids; unicellular cyanobacteria with diameters of a few micrometers pass through many filters.

To minimize analytical errors, the algal cells have to be washed with buffer or distilled water (not applicable with marine or halophilic algae because bursting, or plasmolysis, of the algae will occur) to remove salts and other contaminants. Care has to be taken that the culture is mixed well before taking the samples (duplicates are recommended) and that no cells are lost during separation due to non-sedimenting gas-vacuolated cells.

The collected algae should be dried by a method avoiding excess heat, thereby assuring good reproducibility and identical weights obtained for the same given sample read at hourly intervals. The temperatures commonly applied are between 80 °C and 110 °C. The samples have to be weighed quickly after drying and cooling to avoid absorption of moisture.

Packed cell volume (PCV)

A faster, although crude, method for estimating the amount of suspended solids in a sample is by measuring its packed cell volume (PCV). This is done in special centrifuge tubes of 5–10 ml volume with the bottom extended as a graduated capillary. After spinning the sample (10 min at 1000 g) the solids will settle in the graduated part of the tube, where the amount of the packed cell volume can be read and expressed as a percentage of the total sample volume.

Chlorophyll content

As already mentioned, dry-weight estimations do not exclusively monitor the amount of algae because bacteria and zooplankton may add to the biomass. Since, however, only algae contain chlorophyll, the estimation of this pigment is a reliable, though elaborate, method in algal biomass computation.

Depending on the algal strain examined, acetone (sometimes 80% acetone gives better results than 100% acetone), ethanol or diethylether are used to extract the pigments from the separated algal cells.

In some cases brief heating is required to achieve complete pigment extraction. After that the cell debris is removed by centrifugation or filtration; the extract should be protected from light to avoid bleaching of the pigments.

The following steps are recommended for preparing the chlorophyll extract.

1) Centrifuge 10 ml of algal suspension at 3000 rpm for 1 min and discard the supernatant. For *Spirulina*, centrifugation may not always produce a pellet; sometimes the algae will even float on the surface. In this case, filtration of a given volume through a wide-pore paper filter is the best method for separating the alga from the medium.

2) Suspend the cells in 2–3 ml of the solvent. For ethanol, heating the suspension for about 5 min in a water bath will help to remove

completely the chlorophyll from the cells. Cool the sample to room temperature.

3) Make the volume up to 5 ml. The final volume depends on the actual chlorophyll concentration. If the pigment extract is too concentrated, further dilution is necessary until the absorbance can be read.

The chlorophyll concentration in the extract is calculated by reading the absorption (A) of the pigment extract in a spectrophotometer at the given wavelength against a solvent blank by using the following equations.

For acetone (mg l^{-1}):
Chlorophyll a $= (12.7 \times A_{663}) - (2.69 \times A_{645})$
Chlorophyll b $= (22.9 \times A_{645}) - (4.64 \times A_{663})$
Chlorophyll $a + b = (8.02 \times A_{663}) + (20.2 \times A_{645})$.

For 90% methanol (mg l^{-1}):
Chlorophyll a $= (16.5 \times A_{665}) - (8.3 \times A_{650})$
Chlorophyll b $= (33.8 \times A_{650}) - (12.5 \times A_{665})$
Chlorophyll $a + b = (4.0 \times A_{665}) + (25.5 \times A_{650})$.

For diethyl ether (mg l^{-1}):
Chlorophyll a $= (9.92 \times A_{660}) - (0.77 \times A_{642.5})$
Chlorophyll b $= (17.6 \times A_{642.5}) - (2.18 \times A_{660})$
Chlorophyll $a + b = (7.12 \times A_{660}) + (16.8 \times A_{642.5})$.

Content of water-soluble pigments

Cyanobacteria contain certain water-soluble pigments (phycocyanin and phycoerythrin) that cannot be found in any other algal species. Therefore the occurrence of these pigments in algal extracts indicates the presence of cyanobacteria. Because the appearance of these forms is not desired in cultures of *Chlorella*, *Scenedesmus* or other green algae, the determination of the water-soluble pigments is a qualitative and quantitative method for detecting the presence and estimating the amount of cyanobacteria, especially in those cases where a microscopic control is difficult. For the quantitative estimation of the pigments phycoerythrocyanin (PE), allophycocyanin (APC) and C-phycocyanin (PC), the algal sample is treated with 0.15 M NaCl and the cells disrupted by sonication. After separation by centrifugation, the absorbance of the crude pigment extract is measured at 565, 620 and 650 nm. The pigment concentration is calculated by the following equations (mg l^{-1}):

$$\text{C-phycocyanin (PC)} = \frac{A_{650} - 0.7 \times A_{650}}{7.38}$$

$$\text{Allophycocyanin (APC)} = \frac{A_{650} - 0.19 \times A_{620}}{5.65}$$

$$\text{Phycoerythrocyanin (PE)} = \frac{A_{565} - 2.8 \times (\text{PC}) - 1.34 \times (\text{APC})}{12.7}.$$

Protein estimation

Different methods are available for the estimation of protein. In the following, three procedures that are commonly used for the estimation of algal protein will be described briefly. All methods require the preparation of a standard curve with known protein concentrations (beef serum albumin).

To prepare the algal sample for the protein estimation, the following steps are recommended.

1) Centrifuge 50 ml of sample at about 3000 rpm for 1–2 min.
2) Wash the cells with distilled water (or buffer) and spin again at 3000 rpm for 2 min.
3) Extract the chlorophyll from the algal cells (see methods for chlorophyll extraction on p. 58).
4) Centrifuge sample and discard the chlorophyll-containing supernatant.
5) Resuspend the algal pellet in a small amount of water and then proceed according to the recipes for the protein estimation.

Biuret reaction

The Biuret reaction was one of the first colorimetric methods developed for the estimation of proteins; its color complex has not yet been fully elucidated (Gornall, Bardawill & David, 1949).

Biuret reagent. 1.5 g $CuSO_4.5H_2O$ and 6.0 g NaK-tartrate are dissolved in 500 ml of distilled water. With vigorous stirring, 300 ml of 10% NaOH (w/v) are added, and the volume is made up to 1000 ml with water. The solution should be stored in a dark bottle.

Protein estimation. To 1 ml of protein solution, 4 ml of Biuret reagent have to be added and mixed thoroughly. The sample is kept for about 20 min at 37 °C and read at 540 nm against a reagent blank.

The linearity of the assay lies between 2 and 15 mg protein/ml^{-1}. The Biuret reaction is subject to interference by Tris, ammonia and glycerol.

Estimation after Lowry et al. (1951)

Reagent A. 100 g Na_2CO_3 are dissolved in 1000 ml 0.5 N NaOH.
Reagent B. 1 g $CuSO_4.5H_2O$ are dissolved in 100 ml of water.
Reagent C. 2 g K-tartrate are dissolved in 100 ml water.
Reagent D (Folin–Ciocalteus reagent). 10 g $Na_2WO_4.2H_2O$ and Na_2MoO_4

are dissolved in a small amount of water: subsequently 5 ml 85% phosphoric acid and 10 ml 36% HCl are added. The mixture has to be boiled for 10 h under reflux. After cooling, 15 g LiCl and 1 drop of bromine are added. The solution is boiled for 15 min, cooled and filtered. The volume is made up to 100 ml. Because the preparation of this solution is elaborate, it is recommended to use commercially available reagent.

Reagent E. 15 ml reagent A, 0.75 ml reagent B and 0.75 ml reagent C are mixed together.

Estimation. 1 ml of reagent E is added to 1 ml of the protein sample; the mixture is kept for 15 min at room temperature. Then 5 ml of reagent D are diluted with 50 ml of water, 3 ml of the solution are added to the protein sample and mixed immediately. The samples are incubated for 45 min at room temperature. The absorption of the color complex is measured at 540 nm against a reagent blank. The linearity of the assay ranges from 10 to 60 μg protein/ml^{-1} of the final assay solution.

The procedure is subject to interference by potassium ions, magnesium ions, EDTA, Tris, thiol reagents and carbohydrates.

Estimation after Bradford (1976)

Reagent. 100 mg Coomassie Brilliant Blue G-250 are dissolved in 50 ml 95% ethanol. To this solution 100 ml 85% (w/v) phosphoric acid is added. The resulting solution is diluted with water to a final volume of 1000 ml.

Estimation. 5 ml of reagent are added to 0.1 ml of protein sample and mixed. The absorbance at 595 nm is measured after 2 min and before 1 h. The sensitivity of the test ranges from 5 to 25 μg protein/ml^{-1} in the final assay volume.

There is some interference from strongly alkaline buffering agents.

Microscopic examination

This method is often used for algal quantification. Different models of counting chambers are available commercially, the choice depending on algal cell size and concentration. Because reproducibility of the counts is the main problem involved in this procedure, special attention should be given to sampling, dilution of the medium and filling the chamber. However, even under optimal care the results are often misleading for various reasons.

a) The volume analyzed in a commonly used counting chamber (hematocytometer) is about 1 μl. This volume, and therefore the number of cells in it, is extremely small and does not always represent an average of the algal sample, even if the suspension is mixed properly and attention has been given to sampling, diluting and filling the chamber.

b) The method becomes more questionable if algal strains that grow in the form of coenobia or colonies have to be counted as well as individual single cells, which is the case for instance with *Scenedesmus* or *Micractinium*.

c) Because of the morphology of certain algal species (*Micractinium* with its long spines) the algae will be squeezed off when the cover glass is pressed on the counting chamber, resulting in counts that are too low. For these reasons microscopic examinations should be used more as criteria for qualitative estimations rather than for the quantitative measurement of algal growth.

A comparison of different methods for estimating the biomass and growth of planktonic algae was reported by Butterwick, Heaney & Talling (1982) using the colonial diatom *Asterionella formosa* as a test organism. The methods tested included microscopy and electronic counting, nephelometry, *in vivo* absorbance, *in vivo* chlorophyll *a* fluorescence, reducing capacity (carbon equivalent), spectrophotometric measurement of extracted chlorophyll *a*, and fluorometric estimation of chlorophyll *a*. The main criteria of this comparison were precision, sensitivity, limit of detection and time requirement.

Most methods yielded an acceptable precision over wide ranges of cell concentration; for visual counting and nephelometry, however, the variation exceeded 10%. *In vivo* estimations, although the most rapid, were unsuitable for low algal concentrations. The chemical methods required larger quantities of sample preparation; they were relatively slow but allowed batches of samples to be processed together and hence gave more generalized measures of algal biomass. Whatever method is adopted, some degree of compromise has to be accepted regarding specificity, manipulative ease, precision and limit of detection. To obtain the most reliable results, it is recommended to apply several methods simultaneously, including qualitative microscopical observations to avoid fundamental errors such as change of the organism tested through contamination.

10   Large-scale cultivation

10.1   Pond design

General introduction

One of the important requisites for successful outdoor algal cultivation is the construction of suitable basins, which should be efficient, easy to operate, durable and cheap. Therefore, size, shape, the material used for construction and the type of agitation employed vary depending on the local conditions, the raw material available and the final utilization of the algal biomass. In practice, the construction is always a compromise between high-performance systems with optimal hydrodynamic properties and the economical need to keep the construction costs within acceptable limits.

During the short history of algae mass cultivation, different kinds of ponds have been developed and operated at experimental, pilot and industrial scale.

Two major approaches to the design of such plants have emerged from this work, i.e.

i) sophisticated closed bioreactors for the cultivation of specialized algal strains for the production of specific biochemicals such as enzymes, cytochromes, toxins, radioactive labelled compounds, pharmaceuticals, etc.; and

ii) cheap open units which are easy to maintain in order to compete with other methods for the production of either proteinaceous matter or cell constituents.

Three different principles of open algal ponds design have been developed from the various concepts proposed (Fig. 10.1).

i) Circular ponds with agitation provided by a rotating arm.

ii) Oblong forms (so-called raceways), which are constructed either as a single unit or in a joint form of several units (meander), with agitation by means of paddle wheel(s), propeller, air lift pumps.

iii) Sloping, often meander-like constructions, where mixing of the algal suspension is achieved by pumping and gravity flow.

At some places, closed circular systems have been tested, in which the

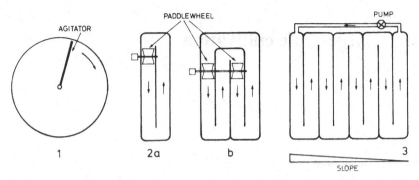

Fig. 10.1. Schematic outline of major algal pond designs. 1, Circular pond with rotating agitator; 2a, single oblong pond ('raceway') with paddle wheel; 2b, joined oblong ponds ('meander') with paddle wheels; 3, sloped meandering pond with circulating pump.

culture is circulated either in plastic tubes or in shallow trenches covered with a plastic ceiling.

Pond-wall material

One of the most important factors that determines the performance and durability of an algal pond is the material used for the construction of the bottom and walls, and the type of surface lining. The specific selection depends on various factors such as purpose of algal production, climatic conditions, availability of material, expertise and equipment for the installation.

The materials used for the erection of the pond side walls range from simple sand or clay embankments, brick or cement constructions, up to more expensive plastics such as UV-light-resistant PVC glass fiber or polyurethane. In most cases the material employed for the construction of the pond bottom is identical to the one used for the wall; however, there are examples where bottom and walls are made from different materials, as will be explained below.

Of special significance is the pond lining material because it is this part which is in permanent contact with the medium and determines seepage, erosion and turbidity and may also affect the chemical composition and quality of the algal biomass produced. In general, two types of lining material can be identified, i.e. membrane and non-membrane.

Nowadays, most microlalgae production systems utilize impervious liners, usually plastic membranes. These membrane linings include the various types of plastic and rubber sheets available today that are often used in agriculture or for the lining of swimming pools, water reservoirs, disposal sites, etc. The material used for algal-pond construction should be durable and uniform, it should prevent loss of media and entrainment of dirt during mixing, be resistant to chemicals and UV light, be tough but elastic at the various temperatures encountered, non-toxic, and easy to seam. Except for very small ponds, plastic sheets require joining and sealing on the spot.

The utilization of large plastic sheets requires engineering experience: large gas bubbles under the bottom of plastic-lined ponds are frequently observed because of leakage under the cover and fermentation of organic matter; tearing of the liners, wrapping around paddle wheels or wind damages are other problems encountered.

At present, plastic sheathings made from UV-resistant polyvinylchloride (PVC) or white reinforced UV-resistant polyethylene sheets are the most common materials used for lining algal ponds. Cost of liners varies from about US\$1 to 10 m^{-2}. The use of cheap material (thin PVC) is not recommended since it does not last very long and the lower initial costs will be offset by early replacement costs, not to mention the installation costs. For lined ponds with a long-life durable liner, costs of about US\$10 m^{-2} have to be calculated; however, costs up to US\$25 m^{-2} have been reported (Benemann, 1986).

The choice between lined and unlined ponds must be based on both technical and economical considerations. Unlined ponds may suffer from silt suspension and percolation, and the inability to be decontaminated after contamination with undesired microorganisms. The non-membrane materials range from compressed soil or clay (which are used in several salt ponds as, for instance, in Australia) to mortar, concrete, asphalt or glass-fiber-reinforced polyester.

Concrete has been used for many years in various algal plants. Although it requires high initial investments, it lasts for many years and hence in the long run is more economic than cheaper, alternative materials. However, for certain purposes, an untreated cement finishing cannot be recommended, as for instance in the case of *Dunaliella* cultivation. The high salt content of the *Dunaliella* medium needs special care for the preparation of the inner pond surface in order to avoid deterioration of the concrete and corrosion of the metal that may come into contact with the medium. As reported by Ben-Amotz & Avron (1989), two of the pioneers in *Dunaliella* cultivation, asphalt concrete or spraying is not recommended, because this material gradually integrates and starts floating as small flocs in the medium. On the other hand, the same material remained intact in other algal ponds run with fresh-water cultures.

All materials will have advantages and disadvantages that affect their suitability. Often, a combination of different materials will provide the most suitable solution to meet the needs at different areas of the pond.

Closed tubular systems

As already mentioned, closed bioreactors are required for special purposes. The high value of certain algal products and the need for good manufacturing practices as well as the necessity for sterile reactors, or at least controlled unialgal reactors, justify the operation of such installations.

The advantages attributed to small closed systems, for instance the prevention of contamination, cannot be transferred to large systems. It has repeatedly been reported that this is not possible, as could be foreseen. Also the facility of better supply and utilization of CO_2 as attributed to covered ponds has not been verified yet, probably mainly because of the fact that CO_2 transfer rates are not large and that the high investment costs for a cover are not warranted. The only advantage of larger covered systems seems to be the possibility of maintaining relatively high temperatures at night, which can become a limiting factor of algae production at several locations (however, it should be mentioned here that there is strong evidence that the maximum increase in algal biomass by providing increased night temperatures seems to be less than 20%, a questionable benefit when compared with the costs for a cover), and to a lesser extent the extension of the cultivation period during seasons with low ambient temperatures.

As must be clear by now, microalgae cultivation is uniquely difficult to scale up in such controlled systems as light limits the design options to high surface area to volume reactors. The design and operation of a few different tubular cultivation systems will be described here with the intention of illustrating the general principles of such devices.

At the Institute for Applied Research in Beer-Sheva (Israel), very simple, although durable and efficient cultivation containers have been developed, mainly for the growth of *Porphyridium*. As can be seen from Fig. 10.2, they consist of polyethylene sleeves of the kind used in aquaculture systems for production of larvae. The bottom of each sleeve is sealed by hot-welding into a cone shape to prevent settling. The culture is mixed by blowing an air stream containing CO_2 – if required – from the bottom into the culture. Excessive heating on hot days was prevented by spraying the sleeves with water when the temperature rose above a certain point (31 °C). The volume of these algae containers is 25–30 l.

In 1950, small photobioreactors (10 l) consisting of vertical glass columns were tested in California. Because it was assumed that insufficient artificial illumination is the major growth-limiting factor of such system, comparable studies were performed with cultures grown indoors and outdoors. It was found that despite the much higher light intensity, the values for the outdoor cultures were not higher than those obtained from cultures growing with artificial light. This can be explained by the fact that in the vertical position the reactor is always at a large angle (A) to the sun's

Fig. 10.2. Outdoor cultivation of *Porphyridium* sp. in polyethylene sleeves of capacity 25 l. (Photograph courtesy of Dr S. Arad, Beer-Sheva, Israel.)

rays, for which a substantial amount of solar energy (I_0) is expected to be reflected, and only a fraction ($I_0 \sin A$) is available for biomass growth.

The first attempts to cultivate algae in an outdoor tubular system were performed by the A.D. Little Company on the roof of their building in 1950. The design suffered from several drawbacks, i.e. leaks in the plastic material used, algal settling and, especially, overheating of the medium, which adversely affects the growth of the majority of photoautotrophic microalgae, a problem central to any closed system. The advantages of closed systems over conventional open ponds are that they can be erected over any open space, can operate at high biomass concentrations, prevent the loss of water by evaporation, and keep out atmospheric contaminations.

The first larger tubular bioreactors for algal cultures were reported in the early 1950s, and consisted of about 12 m of glass or plastic tubes. The volumetric ratio of the tubing system was 0.54. The highest daily growth rates obtained for *Chlorella* was about 12 g m^{-2}.

Another tubular loop 40 l bioreactor was constructed at about the same time in Japan, having a total length of 33 m. It was observed that in a light-limited culture, the linear growth rate (dn/dt) of a *Chlorella* culture (where n

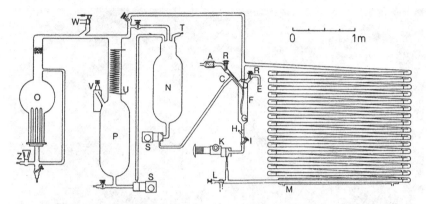

Fig. 10.3. A 110 l tube plant for sterile production of microalgae. A, filter for
CO_2/air mixture; C, entrance port; D, baffle plate; E, outlet vent; F, gas-
exchange cylinder; G, perlator; H, stainless steel tube with rubber septum; I,
valve to direct steam flow; K, centrifugal pump with gear; L, corner valve; M,
water-jacketed tube; N, medium supply vessel (100 l); O, distilling vessel and
steam generator; P, supply vessel for distilled water (150 l); R, valve; S,
centrifuge; T, inlet for new sterile medium; U, cooling coil; V, heated air filter;
W, safety valve; Z, automatic valve for deionized water.

is the number of cells and t is the time) decreased with increasing tubing
diameter from 0.7 to 6.0 cm following the relationship $dv/dt = kd$, where
$k =$ growth rate per unit area and $d =$ tubing diameter.

Two different types of such constructions, holding 30 and 110 l of culture
medium, which have been designed in the author's laboratory, shall be
described here briefly (Jüttner, 1982). The 110 l unit actually represents an
extremely lengthened glass tube with an inner diameter of 40 mm and a total
length of 79 m (Fig. 10.3). The individual tubes are mounted in such a way
that full light is obtained from both sides. The use of tubes with a larger
inner diameter may be possible for organisms with low light requirements
but for many algae this would result in significant decrease of the growth
rate. The algal suspension is circulated with a glass centrifugal pump, the
flow rate of the suspension varies between 20 and 60 cm s^{-1}.

A high flow rate is necessary to prevent settling of the algae; however, at
such high flow rates shearing forces may damage the algal cells, especially
filamentous cyanobacteria. The gas exchange, necessary to supply the
organisms with CO_2 and to eliminate the formed oxygen, takes place in the
gas-exchange cylinder. The CO_2 utilization is very efficient as the inflowing
algal suspension is dispersed with a baffle plate and flows down the wall of
the cylinder in a thin film in which rapid gas exchange can occur.

The temperature is adjusted with a water-jacketed 2 m tube; maximum
temperatures of 38 °C could be achieved with a water bath temperature of
60 °C. Illumination is provided from each side with two adjustable panels
equipped with 60 fluorescent tubes corresponding to approximately 10 000

Fig. 10.4. Small tubular thermostated bioreactor, made from glass pipes (inner diameter 40 mm, wall thickness 7 mm). (Photograph courtesy of Dr Pulz, Berlin, Germany.)

lux from each side. Control of growth parameters can be performed with a by-pass. The whole system can be sterilized with live steam at a pressure of 0.25 atm.

A similar but smaller design (40 l) for laboratory tests has been in operation at the institute for cereal research in Berlin (Fig. 10.4). The system is temperature controlled by means of an external heat exchanger. The performance is registered by pH, temperature and density sensors; the velocity of the culture is regulated by a stepless operating pump.

Many of the techniques employed for the horizontal tube plant can also be applied for vertical tower-type cultivation devices. The main reason for the design of the tower plant was the need for a more economical construction that would permit the cultivation of medium amounts of microalgae. The main part of the system is a glass tube 2 m long and with an inner diameter of 15 cm, with an inlet at the bottom and outlet at the top (Fig. 10.5). To improve the unfavourable geometry of the tower for illumination, a thinner glass tube ('finger') has been inserted from above, eliminating the central space, which is poorly illuminated in dense cultures. In the reduced volume of the tower a higher total production of algae was obtained than in the unreduced tower. The optimum width of the algal suspension depends on the algal species cultivated and the amount of irradiation that can be provided. The widths tested varied between 2.5 and 5

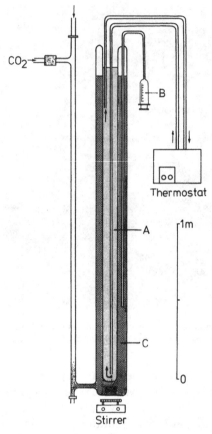

Fig. 10.5. A 300 l tower plant for axenic cultivation. A, Thermostatically controlled finger; B, syringe for withdrawal of samples; C, algal suspension.

cm. If the inner part is filled with water circulating through a thermostat, it can be used to control the temperature of the system. The gas, which is introduced through a laterally inserted connection at the bottom of the tower, keeps the algae in suspension. In order to avoid algae with higher specific density accumulating in certain areas of the bottom region, it was found useful to inject the gas stream not through the centre of the bottom but from the side and to stir the bottom continuously with a magnetic stirrer.

In recent years, improved designs of tubular photobioreactors have been proposed and realized in pilot plants in France, Italy and the USA. Various parameters have been studied, such as turbulent flow, energy requirement for circulation, oxygen and CO_2 balance. Closed culture devices of up to 100 m^2 have been constructed with the aim of producing valuable extracellular algal products such as polysaccharides from *Porphyridium*.

Fig. 10.6. General view of a tubular photobioreactor. Tubes are made of polymethyl methacrylate having 13 cm inner diameter and 4 mm thickness. Volume is about 100 l per square meter of illuminated surface. (Photograph courtesy of Dr Materassi, Florence, Italy.)

The pilot plant constructed in Florence (Italy) by Torzillo *et al.* (1986) for growing *Spirulina* sp. is shown in Fig. 10.6. The reactor is made from transparent flexible polyethylene tubes (14 cm diameter and 0.3 mm thickness) and covers a surface area of 80 m² (8000 l). The maximal length of the circuit was 500 m. Owing to their inadequate mechanical strength, the polyethylene tubes were replaced by Plexiglass tubes having 13 cm inner diameter and 4 mm thickness. These sizes were selected to obtain a surface to volume ratio similar to that of open ponds (100 l per square meter of surface).

The tubes are joined by PVC bends, each connection incorporating a small tube for oxygen degassing. The culture falls into a receiving tank from where it is pumped into a feeding tank. From here the medium is discharged intermittently into the tubes creating a flow rate of 25 cm s^{-1}.

A 100 m² tubular reactor made of polyethylene tubes (diameter 6.4 cm) was constructed in France (Gudin & Chaumont, 1983) consisting of five identical units of 20 m² (Fig. 10.7). Each of them can be separately isolated, which allows inoculation of all or only part of the reactor. The culture is circulated in each tube and each 20 m² unit is equipped with a carbonation

Fig. 10.7. Tubular algal photobioreactor (100 m²) at Cadarache. The unit consists of five solar receptors, floating in a large pond. (Photograph courtesy of Dr Gudin, Cadarache, France.)

column where an air/CO_2 mixture is injected through a sintered glass. The unit is equipped to operate in continuous culture with a constant dilution rate that may be chosen so as to produce preferentially polysaccharides (*Porphyridium* sp.), polyunsaturated fatty acids or phycobiliproteins (*Spirulina* sp.).

Several severe limitations still exist for such tubular systems. The long pathways of some of these reactors result in significant hydraulic problems as well as depletion of CO_2 in the culture. Another major limiting factor is the accumulation of photosynthetically produced oxygen in the tube, which can be demonstrated by a short calculation: during peak hours of photosynthesis the production of 4 g biomass $m^{-1} h^{-1}$ in a tube of 4 cm diameter results in an approximately equal amount of O_2. A 25 mg l^{-1} increase in oxygen tension (= trifold increase of the saturation of 8 mg l^{-1}) would require only slightly over 10 min, demonstrating that degassing of oxygen is of great importance (Benemann, 1986).

Comparable studies with tubular reactors with lengths 250 and 500 m showed that the shorter unit gave higher yields. *Spirulina* sp. produced in the longer reactor had a lower protein content, counterbalanced by a higher carbohydrate content. Because no significant differences in the temperature profile and the illumination occurred, these differences must be attributed mostly to the adverse effect of high oxygen concentration on protein synthesis.

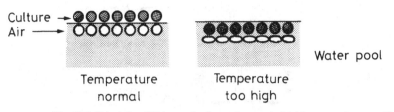

Fig. 10.8. Principle of lifting and submerging a tubular bioreactor in a pond in order to regulate the temperature of the algal suspension.

The tubular reactors are advantageous in areas with moderate temperatures but become problematic in warm climates because the temperature in the tubes may reach values 10–15 °C higher than the ambient air temperature. Therefore it is necessary to devise a cooling system. Several different possibilities have been tested to prevent overheating of the algal suspension.

a) Shading of the tubes with dark-colored sheets. However, to achieve a significant shading effect, up to 80% of the surface area has to be covered, which causes a large reduction of the illumination and consequently in the yield of biomass.

b) Cooling of the culture by spraying water on the surface. Experiments performed in Italy maintaining the temperature of the algal suspension between 33 and 35 °C demonstrated that the amount of water lost by evaporation ranges between 1 and 2 $l \, d^{-1} \, m^{-2}$.

c) Floating the tubular system on a large body of water. This solution ensures sufficient heat exchange besides providing agitation through wave action and reduction of support structures.

With the French bioreactor, thermal regulation is ensured by flexible tubes mounted beyond the cultivation tubes, which may be inflated with air so that the culture tubes are lifted during periods of tolerable temperatures or are immersed when the temperature in the culture tubes becomes too warm or too cold (Fig. 10.8).

Other problems encountered with closed reactors are hydraulic problems (headloss), depletion of CO_2 and accumulation of oxygen affecting yield and chemical composition of the biomass.

In most of the culture systems developed so far, the algal cultures are recycled by centrifugal pumps. It was generally believed that cultures of unicellular algae were not sensitive to turbulence and the hydraulic forces of centrifugal pumps. Recently, Pirt (1983) compared physiological parameters of *Chlorella* cultures grown in a tubular loop bioreactor, recycled either by a centrifugal pump, a rotary positive displacement pump, a peristaltic pump or an air-lift system. It was found that the specific growth was higher with the peristaltic pump compared with the other means tested.

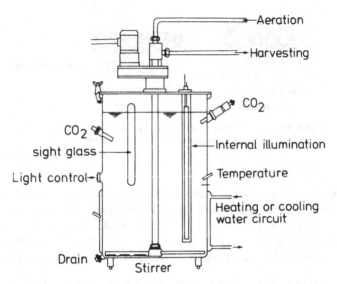

Fig. 10.9. Design of stainless steel photobioreactor. For details see text. (Courtesy of Dr Pohl, Kiel, Germany.)

The adverse effects on cells caused by centrifugal and rotary pumps were proportional to the rotation speed of the pump. It was reported that in the air-lift tubular reactor, consisting of glass tubing of 1 cm bore, a biomass concentration of 20 g l^{-1} could be maintained in continuous-flow cultures. The energy required for recycling the culture was estimated to be 0.6 W m^{-2}.

Cylindrical bioreactors

Representing the various designs published in the literature, a reactor type will be described here that includes most of the features typical for such units (Pohl, Kohlhase & Martin, 1986).

These reactors are made of stainless steel having a cylindrical shape with diameters between 20 and 60 cm and heights between 35 and 100 cm, corresponding to volumes of 20–250 l (Fig. 10.9). On top they are covered by a flat steel lid, the lower part (about one-third) is covered by a double-walled steel cylinder, in which water can be circulated to maintain constant temperatures. Most of the reactors are equipped with an internal illumination system. Either transparent pipes are inserted into the cylinder from the top through holes in the lid, which hold fluorescence tubes, which can easily be removed without opening the reactor, or water-proof fluorescent tubes are inserted directly into the medium.

After inoculation, the algal suspension is agitated by means of a stirrer, in this case a T-shaped multifunctional device consisting of a central tube with two or four blades. Sterile air, enriched with CO_2, is passed via the central tube into the blades from where it penetrates into the medium through capillary holes.

Harvesting the algae, which is usually a highly energy-intensive step, can also be carried out rapidly and efficiently by the above stirrer. When the algae have reached the stationary growth phase, both the aeration and the stirrer are turned off so that the algae settle on the bottom of the reactor. After a certain period of time, depending on the individual sedimentation rate of the specific algal species, the stirrer is connected to a suction pump via its central tube and set into very slow rotation. The thick algal suspension formed at the bottom of the reactor is harvested through suction slits in the blades of the stirrer: this is the principle of the vacuum cleaner. This procedure allows concentration of the total algal biomass of a 250 l reactor into a volume of about 5 l within a few minutes. The reactor remains closed during this process. The harvested slurry can be further concentrated by centrifugation. Fresh medium can be added to the reactor via a sterile filtration device making continuous cultivation possible.

Circular ponds

During the early stages of large-scale algal production, circular cultivation ponds up to 45 m diameter, some of them even covered with glass domes, were operated primarily in Japan and Taiwan (Fig. 10.10). The circular system, however, has several disadvantages, because it requires expensive concrete constructions and high energy consumption for stirring, although the turbulence is low in the central part of the pond. Moreover, there are mechanical problems in controlling the movement of the stirring device and in supplying CO_2 to the culture. Finally, land use is inefficient. Today, circular ponds are used in very few places only; most of the algal production units prefer horizontal 'raceways' ponds.

Oblong ponds

The majority of the ponds employed nowadays are of the oblong type, either in the form of a single unit, like the so-called 'raceway' with a central dividing wall, or as a meandering channel obtained by joining individual raceways.

In the following, several examples of oblong pond designs are given, ranging from small experimental basins up to large production units. The figures illustrate the wide spectrum of possibilities in the construction of algal cultivation facilities and also may serve as suggestions for others who are planning to set up algal cultivation ponds.

The most simple constructions are shown in Figs 10.11 and 10.12.

Fig. 10.10. Circular algal ponds with rotating agitator (Taiwan).

Fig. 10.11. Rural algal pond with polyethylene lining at MRCR, Madras (India).

Fig. 10.12. Simple raceway pond, dug in sand and covered by polyethylene lining at Sede Boker (Israel).

They are no more than a 'hole in the ground' consisting of shallow ditches dug into the ground covered with thin plastic sheets draped up sloping earth berms and fixed with soil-filled trenches alongside the pond or just hanging down the side walls of the ditches fastened with bricks. Although the construction of such ponds is cheap and easy and can be done without much technical skill, it is afflicted with several drawbacks. As can be seen in the examples given, the plastic lining is full of wrinkles, favouring the accumulation of solids and simultaneously making cleaning of the pond very difficult. Furthermore, the thin material is exposed to punctures or other mechanical damage and displacement due to wind and hydraulic forces, especially at bends and near mixers, because wind blowing across the berms has the tendency to lift the membranes at the edges. Bubbles due to gas or water accumulation under the sheets are additional problems encountered with such ponds. The risk of damage to the membrane liners, particularly of the bottom layer, can be reduced by providing a padding as smooth as possible. Good results were achieved with white sand, which is comparatively cheap, easy to handle and available almost everywhere. The sand should be sieved through a fine mesh sieve in order to remove sharp-edged stones and pebbles that might puncture the plastic material from beneath, especially during cleaning operations. For large production ponds it is advisable to line the bends and other areas that are of irregular shape or subjected to heavy hydraulic stress with different materials, for instance membranes with fabric reinforcement. It is a rule to keep such plastic lined-ponds always filled with water, even if they are not being used for algal production. The water protects the material from becoming brittle due to

Fig. 10.13. Experimental pond made from tight durable plastic membrane over styrofoam blocks at Sede Boker (Israel).

heat absorption and intense UV-light irradiation, hence minimizing the formation of cracks; furthermore, it prevents displacement due to wind forces. It has been reported recently that the sheet plastic liner from a 1 ha pond at a commercial farm in California was lost within minutes due to high winds because of insufficient water cover.

Another imperfection of such simple pond constructions is the fact that these ponds lack any side walls reaching above ground level. First of all, digging a pond into the soil requires much more earth-work than erecting walls on the ground and using the levelled soil surface as bottom. Secondly, the risk of pond contamination is inherent in such a design because of insects or amphibia (it is not unusual for frogs to be found swimming happily in the pond . . . until they reach the paddle wheel . . .) and dirt blown into the medium, or even loss of the entire culture due to flooding during heavy rains.

The design shown in Fig. 10.13 is more advanced. The walls of this type of pond can be constructed from plastic blocks, concrete blocks, bricks, or even blocks made from sun-dried clay ('adobe'), all of which are equally suitable, covered with a durable plastic membrane, cut and sealed exactly according to the measurements of the pond. The lining is fixed with a rope tightened through metal eyelets along the outer side of the wall. The vertical free-standing wall of this construction minimizes the tendency for wave run-up observed along sloping banks, reduces the space requirement, and permits common side walls between joined units. The small algal growth basins shown in Fig. 10.14*a–d* do not need further explanation but should simply illustrate the wealth of ideas of the researchers working with algae, and may serve as stimulus and suggestion for others, looking for the design of simple experimental algal ponds.

The small unit operated in Singapore (Fig. 10.15) shows some technical shortcomings, i.e. the paddle wheel is oversized, the culture is too deep and hence light limited, and the unprotected walls absorb considerable amounts of heat, which may lead to overheating of the culture.

Somewhat different are the foundations of the ponds given in Figs 10.16 and 10.17. The wall of the pond shown in Fig. 10.16 consists of prefabricated parts of triangular cross-section 5 m in length made from polyurethane foam and lined with a UV-resistant plastic sheet. The parts are joined and sealed with strips of the plastic material glued to the joining ends. This concept offers the possibility of varying the length and width of the pond by adding or removing parts of the wall, which can easily be cut with a sharp knife or a saw. The bottom layer is made by rolling webs of the plastic sheet on a sand layer over the full length of the pond and sealing it to the side wall.

The frame of the pond given in Fig. 10.17 is made from straight and curved metal pipes (1 inch diameter) joined together and resting on small metal supports, as can be seen on the left and right sides of the front part. The plastic lining, reinforced with nylon fabric, is ready-made according to the measurements of the pond, and fastened with a rope pulled through eyelets.

Whereas the small concrete ponds shown above can be constructed without difficulties on native soil, larger ponds require a certain amount of foundation work. To avoid cracking of the concrete bottom layer due to temperature changes and movements of the soil, reinforcement with iron grids is recommended as shown in Fig. 10.18.

Probably the largest algal ponds lined with plastic liners can be found in California. Fig. 10.19 shows part of the *Spirulina* production unit of the Earthrise Company, which consists of 10 production ponds each of 5000 m^2 and a series of research ponds of 1000 m^2, 200 m^2 and 50 m^2, respectively. The depth of the ponds varies between 30 and 40 cm.

Cement, mortar and bricks may cause some difficulties if they come into contact with the culture medium, because there is the possibility of calcium and other minerals leaching out from the material, which might be accumulated by the algal biomass. Such an effect was observed in Thailand for instance, where an unfavourable Ca:P ratio in the algae, caused by the cement plastering of the pond, entailed teeth anomalies in rats fed this algal biomass. The sophisticated algal production units operating in the USA contrast sharply with the simple algal cultivation ditches found for instance in India. Here, several different types of smaller ponds constructed from bricks, mortar or cement are in operation for the production of *Spirulina*. The unit shown in Fig. 10.20 includes several ponds, 10 m long and 1 m wide

Fig. 10.14. Small experimental ponds (*a* and *b*, Dortmund, Germany; *c*, Technion-Haifa, Israel; *d*, Bloemfontein, South Africa).

with a depth of 40 cm (volume 2500 l). There is no partition in the ponds and no baffles as agitation is done manually with brooms or brushes.

The elliptical pond with a total area of 95 m² (12 000 l) was constructed with the intention of harvesting the algal biomass by flocculation (see top left of Fig. 10.20). The pond is divided into four parts: a central sump pit, a central storage section, an inner and an outer culture recirculation section. Each part is separated by a wall 30 cm high with several holes that can be closed by plugs. The inner part is provided for collecting a certain amount

c

d

of the medium, which is pumped through a pipe from here to the outer part. The storage section can be used for retaining the total volume if necessary and also for collecting the supernatant of the medium after harvesting the algae by flocculation or sedimentation. The floor level of the third section has a gradual slope. This is intended for removing the supernatant by opening holes situated at a certain height in the separation walls after settling of the culture. A pump delivers the medium through a discharge pipe from the inner section to the outer one, from where the suspension flows back by gravity in a thin layer. The flow velocity can be regulated by the number of holes opened or closed in the inner walls.

It has already been mentioned that under certain circumstances covered

Fig. 10.15. Small experimental pond made from bricks, showing poor design with disproportionally high side walls and oversized paddle wheel (Primary Production Department, Singapore).

Fig. 10.16. Algal cultivation pond made of PVC sheets sealed on polyurethane blocks (CFTRI, Mysore, India).

Fig. 10.17. Algal cultivation pond (11 m²) made of PVC-coated nylon fabric stretched over an iron frame (CFTRI, Mysore, India).

ponds may offer some advantages compared with open systems. The major advantage is the extension of the cultivation season in areas with temperate climatic conditions. Several of these covered cement ponds, each having a surface area of 500 m², have been constructed near Nanchang (People's Republic of China). This greenhouse-like system consists of cement ponds, equipped with paddle wheels, roofed by iron frames covered by transparent plastic sheets (Fig. 10.21). Besides problems of overheating during hot summer days, it remains to be seen whether the additional construction costs can be compensated for by the extended cultivation period.

Fig. 10.18. Construction of algal pond from bricks and mortar with iron grids for reinforcement of bottom layer (Kasetsart University, Thailand).

Fig. 10.19. Overview of a series of raceway ponds (total area about 100 000 m²) of the Earthrise farms in Southern California (USA) made from plastic sheets.

Fig. 10.20. Different algal ponds made from bricks and cement for rural cultivation of *Spirulina* (CFTRI, Mysore, India).

Fig. 10.21. Raceway cement tank, which can be covered with plastic sheets during periods with unsuitable climatic conditions (Nanchang, People's Republic of China).

Fig. 10.22. Algal cultivation unit at Auroville (India); for details see text.

A quite unusual algal cultivation unit was installed in Auroville, India (Fig. 10.22). In this system, the algal culture is pumped to the rooftops of the buildings standing around the pond from where it descends in a thin sheet and then goes into the narrow outside ring of the basin. Several ports allow the suspension to enter the second ring where normally the water level is high enough for the culture to pass over the inside wall into the center section. Here the culture again is very shallow, allowing full penetration of the light. At the very center there is a deep well that collects the culture, and from where it is pumped into an outside tank where CO_2 from composting and yeast culture units is added – and then back to the roof. During the night, the pump is switched off and the culture is kept in the pond.

For the production of algal biomass that is not designated for human consumption, and hence can be grown in less-clean systems, a more simple pond construction can be used. Such a unit, designed for sewage treatment and production of feed-grade algae, has been set up in Singapore (Dodd, 1986). Here, a rolled crushed rock bottom lining was used over clay without any further stabilization. The walls were made from vertical free-standing corrugated asbestos–cement roofing panels set into shallow concrete trenches and sealed with silicon sealant between the panel joints (Fig. 10.23). This material proved to be very suitable both for straight and curved sections of the wall and can be recommended for similar purposes. Care has to be taken while working with asbestos-containing materials because asbestos fibers have been classified as potentially carcinogenic. If there are concerns about possible contamination of the algal biomass, the inner part of the wall may be covered with a spray-applied membrane. A modification of the joint meandering raceway ponds has been suggested by Märkl & Mather (1985) that permits the partition of one large unit into several

Fig. 10.23. View of algal pond made of corrugated cement sheets (Primary Production Department, Singapore).

smaller ponds, which can be operated individually. The system requires swivelling walls at one end of the smaller sides of the separate channels (Fig. 10.24) and additional mixing devices. The construction is especially useful for scaling-up operations, where a fresh culture can be started in the smallest unit, which then is enlarged stepwise by opening additional channels according to the increase in algae biomass. Fig. 10.25 shows the design of an oblong ('raceway') pond including the presently considered optimal configuration.

Sloping ponds

The principle of the sloping design is to create a turbulent flow while the algal suspension flows through sloping channels or surfaces. A pump returns the algal medium from the lowest point to the top. The turbulence is produced by gravity, according to the flow speed of the liquid given by the slope of the surface. The major advantage of this design is the fact that, except for a pump with a high flow rate and low pressure, no mechanical devices such as paddle wheels or impellers are necessary to achieve mixing of the medium.

The concept of employing sloped cultivation units for outdoor algal mass cultivation was established during the 1960s at Trebon (Czech Republic). In the initial cultivation units, the descending flow surface was set up with shallow troughs made of reinforced polyester resin arranged stepwise one upon another, to form a cascade of hydraulic jumps (Fig.

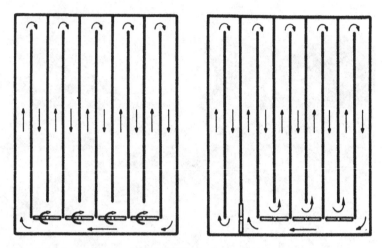

Fig. 10.24. Design for proposed variable algal pond system with swivelling side walls (Märkl & Matern, 1985).

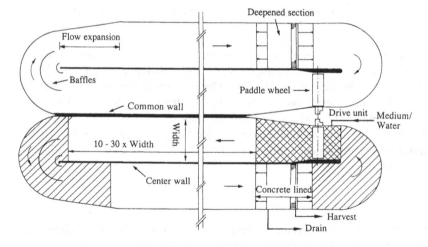

Fig. 10.25. Schematic diagram of an optimal design of a joined 'raceway'-type algal pond.

10.26). The culture was continuously recirculated to the highest trough from the lowest by pumping, so that the medium cascaded by gravity essentially in a thin sheet of liquid. The major drawback of this construction was that the motion in the troughs was not vigorous enough to prevent completely the sedimentation of algae. The relatively high slope needed a considerable amount of energy for pumping and the aeration created by the down flow of the culture caused high rates of CO_2 desorption from the suspension. Thus the design was modified in order to establish a steady and

Fig. 10.26. Prototype of sloped algal pond in which the algal suspension cascades over a series of troughs (Trebon CSFR). The steps are: 1, storage tank which holds total volume of algal suspension; 2, draining trough; 3, trough made of fibre-reinforced polyester; 4, outlet overflow; 5, discharge piping of main circulation line; 6, gas contactor for CO_2 absorption; 7, discharge piping-bypass line; 8, centrifugal circulation pump; 9, suction pipeline. (Photograph courtesy of Dr Kubin, Trebon, Czech Republic.)

uniform flow of a relatively thin layer of suspension. Two units of 50 m² and one of 900 m² were constructed in which the suspension flowed on a slightly inclined plane surface (3%) made of glass sheets, fitted by transversal baffles which created the necessary turbulent motion. A specific feature of this design was the difference in day and night operation modes. The suspension was kept circulating only during those hours of the day when there was sufficient incident radiation. During the other hours the suspension was kept in a large tank where it was aerated and mixed (Setlik, Veladimir & Malek, 1970).

Optimum preconditions for the construction of algal plants are given in localities with high numbers of sunny days per year and where cheap carbon sources, for instance carbonic mineral springs, are available. These preconditions are given at Rupite, a small village in the South-west of Bulgaria. Here, an algal plant of 3000 m² with sloping ponds (500 m², 1000 m² and 1500 m²) has been constructed, based on the principle developed in Trebon. The cultivation units consist of inclined planes (3%) made from concrete and covered with iron grids (Fig. 10.27). The medium is recycled from the lowest point, where it is collected in a large trough, to the highest point by means of pumps; turbulence of the suspension is caused by the iron baffles.

Fig. 10.27. Sloped (3% slope) algal cultivation unit made of concrete.
Turbulence of algal suspension enhanced by iron grids (Rupite Bulgaria).

CO_2 is obtained by degassing the spring water, which has a CO_2 content of 320 mg l^{-1}. The average daily yields (*Chlorella* and *Scenedesmus*) for the period April–October are reported to vary between 17 and 20 g m^{-2}. A modified version of the principle of sloping algal ponds was constructed in Peru (Heussler, 1985). This unit represents an opened raceway, consisting of a sloped meandering channel, adapted to the terrain gradient by meanders (Figs 10.28 and 10.29). Similar to the other systems described above, the algal suspension is pumped from the lowest point to the top, from where it flows back as a thin film. The plant consists of one 500 m² pond and two 1000 m² ponds. The channels are built from self-supporting concrete segments of 5.5 m length, 1 m width and 0.1 m thickness, linked together by joints sealed with plastic strips. Effluent from harvesting is returned to the inlet so that the flow rate in the pond is not altered during harvesting.

A common problem encountered with all types of ponds made from concrete was also observed with this construction. Although the channels were plastered as smoothly as possible, thick crusts developed on the surface, composed mainly of lime, precipitated from the medium, diatoms and filamenteous blue-green algae. Infected *Scenedesmus* cells in particular were attached to these crusts, making control of parasitic infections very troublesome. To overcome these difficulties, all surfaces in contact with the algal suspension had to be painted with epoxide paint. Optimum operation of sloping algal ponds depends on the adjustment of pump capacity to the flow rate of the medium in the channel; however, efficiency of the system will not be affected seriously by deviations from the hydrodynamic balance.

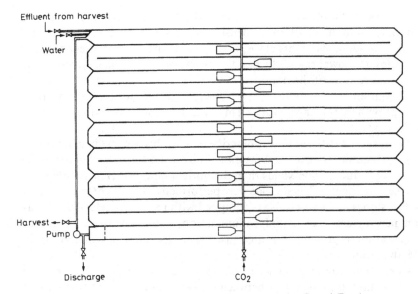

Fig. 10.28. Design of sloped meandering algal pond at Sausal (Peru).

Fig. 10.29. Close-up view of sloping pond at Sausal (Peru).

In the case of an increased flow rate in the channel exceeding the pump capacity, damming up of the suspension will occur at the lower part of the pond. In order to avoid loss of suspension due to these imbalances or failure of pumps, the lower rim of the pond has to be elevated sufficiently to retain the whole volume of the pond.

Maximum yields of 40 g m^{-2} d^{-1} were obtained at times, the average daily growth rates being about 22 g m^{-2} during semi-continuous operation of the plant.

Because data in the literature describe remarkably higher yields obtained with inclined surfaces, it is important to know whether the sloped ponds do actually produce higher yields than the horizontal units. Hence, comparative measurements of algal yields were performed at Florence (Italy) on two baffled slopes and two raceways equipped with paddle wheels. All cultivation parameters (CO_2 supply, illumination, culture density) were kept similar as far as possible. The algae used were *Scenedesmus obliquus*, strain 8M and strain 276–3a.

Consistent differences in performance were found between the two types of cultivation units as well as between the two strains. The productivity of the horizontal ponds was lower than the productivity of the sloped ponds; in the latter the strain 8M was significantly more productive than the strain 276–3a. In the sloped ponds the average daily yields were 24.8 g m^{-2} for strain 8M and 19.5 g m^{-2} fo the other strain, while the yields in the horizontal ponds were 17.2 and 17.5 g m^{-2}, respectively. Better yields from the sloped ponds seem to be a result of their higher temperatures caused by the heat transfer from the circulation pump to the medium. These temperature differences were most pronounced during the morning hours. Because strain 8M is more thermophilic (optimum temperature 35 °C), its higher productivity compared with the other strain may be caused by the very favourable temperatures during the major part of the experimental period. This was confirmed by the observation that, at a later season with temperatures of the medium around 30 °C, the yields of the two strains were comparable.

Hence, no final conclusions can be drawn from these results; however, it seems that the different thermal regimes of the two systems may be the only factor responsible for the differences observed.

Removal of solids

If algal ponds are operated continuously over long periods of time, settleable solids of different origin will accumulate in certain areas of the pond. Removal of such material may be difficult unless certain provisions are made in the pond design. Promising results have been obtained with a modification of the pond construction commonly used that permits the control of solids without interrupting the operation of the pond (Dodd, 1986). The method is based on a depressed section provided in the pond bottom that acts as a sedimentation basin from where the accumulated solids can be removed periodically either through a drain or by means of a hose connected to a displacement pump. The depth of this depressed section should not exceed 1 m to avoid risk of uplift and flotation when the pond is dewatered. As detailed in the section on agitation of algal cultures (p. 95), the mixing velocity normally varies between 5 and 30 cm s^{-1}. In this context it should be considered that at low mixing velocities solids tend to settle at various locations of the pond bottom. Cleaning such slow-mixed ponds is greatly facilitated by periodically increasing the mixing velocity to at least 30 cm s^{-1}, which will move the settled particles on-stream along the bottom until they are captured and accumulated in the deepened section. Brooming may be still required in some cases in order to dislodge compacted heavy settled solids adhering to the pond surface. Where desired, the solids harvested may be thickened by a settler and the supernatant returned to minimize loss of the medium.

Natural ponds

The largest single algal production unit exists in Mexico, established in the early 1970s by the Sosa Texcoco Co. near Mexico City. Here *Spirulina* grows naturally in the remainders of Lake Texcoco, the site of the precolumbian Aztec civilization, located in the valley of Mexico City, at an altitude of 2200 m above sea level in a semi-tropical climate. This lake used to contain large amounts of the filamenteous cyanobacteria *Oscillatoria* and *Spirulina*, both of which were harvested by the Indians and used as food. Today, the alga is cultured in the external ponds of an artificial giant solar evaporator of spiral shape ('caracol', the Spanish word for snail) with a diameter of about 3000 m and a surface area of about 900 ha (Fig. 10.30). As mentioned before, the Aztec Indians collected *Spirulina* from the waters of Lake Texcoco in the early days, long before the arrival of Cortez in 1513. In those days, the lake covered a large area (Fig. 10.31); in the course of time, almost the entire lake dried out and now Mexico City sprawls over what originally was Lake Texcoco, the place from where the oldest reports on the consumption of microalgae originate. The caracol is produced by pumping the waters of the ancient lake to the surface. By a system of dykes the water flows outwards in a spiral path from the point of pumping. By the time the water reaches the end of the spiral most of it has disappeared by evaporation, leaving mostly sodium carbonate and bicarbonate. At a certain part of the spiral, divided from the main section of the pond with wooden baffles, the conditions are favourable for the growth of *Spirulina*. The value of the algae, which formerly were a nuisance to the salt-recovering operations, was recognized in 1967, stimulated by the various reports on the possible commercial exploitation of this type of biomass. The algae are harvested as a kind of 'by-product' by the producer of sodium alkali chemicals that operates a plant to extract soda from the lake. The

Fig. 10.30. Aerial view of spiral algal pond ('caracol') at Lake Texcoco (Mexico).

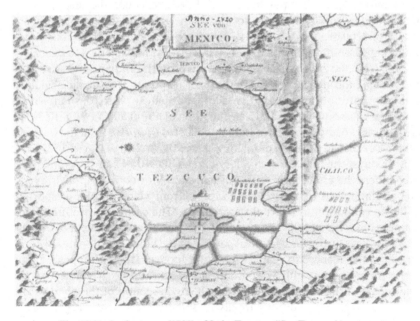

Fig. 10.31. Ancient map (1720) of Lake Texcoco ('See Tezcoco'.)

algal biomass is recovered by filtration under vacuum and then rinsed with water. After homogenization and pasteurization, the algal concentrate is dried with low-heat spray drying. Daily production of the plant has been reported to approach 2 tons of dried algae from about 20 ha of cultivation area, averaging $10 \text{ g m}^{-2} \text{ d}^{-1}$ with peak production rates of $18\text{--}20 \text{ g m}^{-2} \text{ d}^{-1}$ under the most favourable conditions.

The second place where *Spirulina* occurs naturally in such a quantity that a commercial production unit could be established, is situated in an area with old extinct volcanic craters filled with alkaline water, located near the town Butalin in Central Burma. In the large (80 ha) main crater lake, called Twyn Taung Lake, and three other smaller lakes, *Spirulina* grows almost as monoculture throughout the year with peak production during the summer months March and April. During this season, *Spirulina* is found concentrated as thick mats in certain corners of the lake due to wind action. Here the biomass is collected with buckets from boats, filtered first through a coarse screen to eliminate possible fragmental debris and then concentrated by manual filtration by a factor of 10–20. This slurry is further dewatered at the production unit located close to the shore by filtration through polyester cloth bags, which are squeezed out by hand pressure. If the biomass is not collected directly with boats, the *Spirulina*-containing surface water is pumped directly to the plant, where the cyanobacterium is harvested on inclined filters for further processing. After rinsing the concentrate by resuspending it in fresh water, a screw press is used to remove most of the water until a thick paste is obtained, which subsequently is squeezed manually on plastic sheet frames (2 × 2 m in size) by means of small extruders. These frames are set up upright close together slightly inclined in the open air, where the paste dries in the sun within 2–4 h. The dry *Spirulina* flakes are ground twice in a mill into fine powder, which finally is punched into tablets by a pharmaceutical factory at Rangoon.

Since 1988, when the factory began with the pilot-scale production of *Spirulina*, the annual yield was increased continuously up to an amount of about 13 t, produced in 1992.

10.2. Evaporation

A major problem with algal cultivation in dry tropical areas is the high rate of evaporation from the open pond surface (up to $10 \text{ l m}^{-2} \text{ d}^{-1}$). This fact poses a problem both from the point of increasing the salt concentration in the medium and in the aquisition of sufficient water to make up for the water loss. For this reason any location selected for the construction of an algal plant should have an abundant source of fresh or low-salt-content make-up water. On the other hand, certain tropical areas with frequent monsoon rains may have such amounts of rainfall that severe culture dilution and loss of nutrients and algal biomass may occur. For such areas it has been proposed to equip the ponds with overflow spillways and to provide covered deep storage ponds into which the cultures can be pumped temporarily. Besides evaporation losses, the climatic conditions also affect the temperature of the culture medium. With low relative air humidity, high rates of evaporation occur, which have a cooling effect on the medium. At high relative air humidity and no winds, the medium may heat up to as much as 40 °C, which is lethal to almost all algal species. Thus, sites selected for algal cultivation should be located in areas with an average humidity below 60%.

10.3.   Agitation

General considerations

Provided that the environmental conditions are favourable and the supply of nutrients is sufficient, agitation of the culture medium is one of the most important requirements in algal mass cultivation for obtaining consistently high yields of biomass. Agitation implies a combination of various effects, e.g. uniform dispersion of algae to ensure frequent exposure to light and avoidance of algal settling to the bottom of the pond, homogenous distribution of nutrients, better utilization of carbon dioxide, and prevention of thermal stratification. In unmixed shallow ponds, temperature differences of as much as 8 °C can be found from top to bottom on warm days. Such differences arise when the top layer of a dense culture absorbs and converts most of the irradiance to heat, which generally occurs with a ten-fold higher efficiency than the conversion of light to chemically bound energy. Because of its specific density, the warm water remains on the surface and the algae in this layer deplete it of bicarbonate ions so that the pH may rise above 10. Under these conditions essential nutrients precipitate, carrying part of the algae to the bottom. With clearing of the top layer of the medium the next-deepest layer is exposed to light and the same sequence starts again until almost all algae become precipitated on the pond bottom in the form of large agglomerations, which tend to flake off the bottom and rise to the surface where they may become anoxic and odorous. Hence, the primary function of mixing is to keep the algal cells in suspension and periodically expose them to light in order to ensure photoautotrophic growth and to avoid deposited solids. If the mixing velocity is too slow, dead algal cells and other forms of organic debris will accumulate on the bottom of the pond, especially in areas where the turbulence is small. The formation of such stagnant zones will lead to the formation of anaerobic conditions, which effect cell deterioration thus decreasing the yield and impairing the quality of the algal biomass. Under certain circumstances toxic compounds might even be formed from the decomposing organic material, causing the loss of the entire culture.

The need for uniform distribution of minerals in the medium implies the prevention of nutritional and gaseous gradients around the algal cells formed during active metabolism. Such gradients impose restrictions on the growth and are alleviated with sufficient turbulence. Like any photosynthesizing plant cell, algae produce oxygen, which creates high concentrations of dissolved O_2 in the medium. During peak hours of photosynthesis and in a stagnant medium, oxygen supersaturation of more than 400% is quite common. Since this oxygen is the final product of photosynthetic activity, it is written on the right-hand side of the general photosynthesis equation:

$$6CO_2 + 12H_2O \longrightarrow 6(CH_2O) + 6H_2O + 6O_2.$$

As can be seen, in relation to the CO_2 supply, the accumulation of O_2 must also be considered. For each mole of CO_2 fixed about one mole of O_2 is liberated. When the O_2 tension in the culture rises above that at equilibrium with the atmosphere ($7–9$ mg l^{-1}), which commonly happens in healthy cultures during the afternoon hours, when the O_2 tension can be up to four times higher than air saturation, it will cause a shift of the above equation to the left resulting in inhibition of the photosynthetic efficiency, photorespiration and even photooxidative death.

Another important effect achieved by adequate mixing of the culture is a fast movement of the algal cells from the illuminated upper zone (photic zone) of the pond to the lower, dark one and back again to the surface, resulting in a dynamic light–dark pattern for the individual algal cell. The phenomenon of mutual shading, i.e. the continuous change between light and dark phases, represents one of the most basic requirements for high biomass productivity. In a dense culture, this photic zone, where the algae receive sufficient light for photosynthesis, can be very shallow (2–5 cm) so that vigorous mixing is required to provide a uniform average light exposure of all cells. In a pond maintained at the optimal population density, it is estimated that about 85% of the cells do not receive sufficient radiation to facilitate a photosynthesis rate above the compensation point. Most of the mixing techniques used are insufficient to ensure attainment of the maximum photosynthetic potential of the algal cells, because the turbulence induced is random, which means that not all algae are subjected to a uniform pattern of movement into and out of the illuminated zone of the medium. Although in several systems the algae are exposed to varying light intensities due to the turbulence of the medium, this random light exposure is not sufficient to take advantage of the flashing light effect to any significant degree, even if the stirring or pumping is very intense.

Increases in photosynthesis of up to 87% have been reported by simply modulating the irradiance on algal cells on a time scale ranging from minutes to hours. Regardless of the physiological mechanism involved, it seems clear that photosynthetic efficiency may be enhanced by a factor of two or more by modulating or flashing the incident light. For the flashing light it is apparent that the maximum duration of the light pulse that produces the highest efficiency will be negatively correlated with light intensity greater than or equal to that above which photosynthetic efficiency begins to decrease under constant illumination. It has been concluded that the duration of the dark period must lie somewhere in the approximate range of 0.1–10 s.

The mixing velocity, i.e. the flow rate of the medium, necessary to ensure all the effects required for optimal algal growth, varies primarily depending on the settling rate of the specific algal cells. It is assumed that a velocity of 10 cm s^{-1} is generally sufficient to avoid deposition of cells. However, an average value of 20 cm s^{-1} is often used because of unavoidable fluctuations of the velocity, particularly at the bends of the horizontal 'raceway' ponds. In practice, most of the ponds are operated at a speed between 10 and 30 cm s^{-1}.

Unsolvable difficulties in maintaining a constant mixing velocity over the total area of circular ponds was one of the main reasons for discontinuing the employment of this type of pond. Mixing in these constructions was obtained by means of one or two rotating arms fixed in the center of the pond like spokes of a wheel. Since the angular velocity increases with increasing length of the stirrer, speed and turbulence were optimal at the outer parts of the pond but decreased towards the inner sections resulting in almost stagnant water around the center. Several auxiliary constructions were tried to achieve sufficient mixing but all of them were quite susceptible to problems and gave unsatisfactory results.

Apart from this extreme case, all other types of pond operating at average mixing velocities encounter difficulties with sedimentations of algal cells and other solids at particular areas, especially at the curved ends of oblong ponds. Construction details for overcoming these problems are described in the section dealing with pond design (p. 63).

If the length of a straight pond channel exceeds a certain limit, a sloping pond bottom becomes necessary in order to maintain uniform flow rate and constant shallow depth. In addition, when proposing large ponds, the technical difficulties of holding the construction tolerance on very flat gradients to a few centimeters have to be considered. As the current velocity and pond area increase, the head loss at the loops will exceed the culture depth. This may lead to operational problems. If, for instance, the height difference of the pond bottom between both ends of a sloped pond exceeds twice the culture depths, shutdown of the mixing device will cause the medium to collect in the lower area. If this happens, the upper part of the pond will begin to dry out, besides the risk that the medium may overflow the pond if the wall at the lower part is not high enough. Hence, for practical reasons it seems advisable to keep the head loss below twice the culture depth. For these conditions, Dodd (1986) has proposed a model for approximating the correlation between pond area, mixing velocity and culture depth.

The flow velocity of water in a sloping channel is described by the Manning equation, which relates power need to the mixing velocity by considering the friction at surface (roughness coefficient), the hydraulic radius and the slope of the channel:

$$V = (R^{2/3}/n)\,(H^{1/2}/L),$$

where V = flow speed (m s^{-1}), R = hydraulic radius (m) = (quotient of area by wet circumference of the cross section of the channel), n = Manning friction coefficient (s m$^{-1/3}$), H = head loss (m), and L = channel length (m). The friction factor varies according to the relative roughness, which is defined as the mean height of bottom discontinuities compared with the depth of the water in a channel, i.e. a discontinuity of 1 mm in height in a 10 cm deep channel would have a roughness of $0.1/10 = 0.01$. With the help of the Manning's equation it is possible to calculate the maximum size of algae ponds based on assumptions about roughness of pond lining, channel width, mixing velocity and tolerable headloss.

By rearranging the equation and substituting the depth D for R, which holds for very wide, shallow channels, and assuming that the head loss is equal twice the culture depth D, the equation can be solved for L as follows:

$$L = (2 \times D^{7/3})/(n^2 \times V^2).$$

In this model the width of the channel is assumed equal to $L/20$, so that the total pond area becomes $A = L \times L/20$. By substituting L, the formula can be expressed in terms of limiting pond area A:

$$A = 2D^{14/3}/10^5 n^4 V^4.$$

For smooth concrete surfaces, a Manning coefficient of 0.013 can be assumed. For concrete coated with epoxi-paint or tightly spread plastic sheets, the coefficient is lowered to 0.010 (Table 10.1). If this equation is plotted for various channel depths, a family of curves will be obtained (Fig. 10.32). It should be stressed that these curves are not hard and fast rules, but help to show the relationships involved. A major implication in practical pond engineering is that channel-bottom slopes must match the gradient to maintain a reasonably constant channel depth.

As can be seen, the limiting pond area becomes very small for shallow depths and high mixing velocities. On the other hand, by increasing the depth and reducing the velocity, the limiting pond size increases very quickly until a point is reached where the area becomes so large that constructional and operational, rather than hydraulic, problems will limit the pond size. The second important limiting factor is the capacity of the mixing device to produce the necessary lift of the medium required to accomplish on adequate velocity. For the most common method, i.e. mixing by paddle wheel, a total lift of about 30 cm is probably the upper limit.

Experiences with large-scale culture units indicate that a mixing velocity of 5 cm s^{-1} is sufficient to prevent thermal stratification and to keep the

10.1. *Manning friction factors for different materials*

Material	Manning factor
Plastic on smooth concrete	0.008
Plastic on smooth earth	0.010
Plastic on granular earth	0.012
Smooth cement concrete	0.013
Smooth asphalt concrete	0.015
Coarse concrete	0.016
Gunnite or sprayed membranes, smooth earth	0.020
Rolled gravel, coarse asphalt	0.025
Rough earth	0.030

Source: After Oswald (1988).

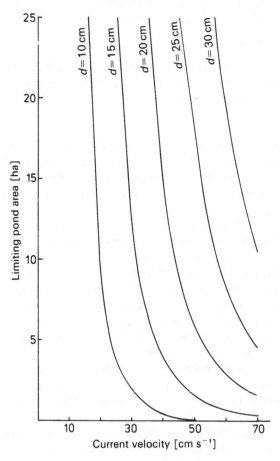

Fig. 10.32. Limiting pond area–depth–velocity relationship.

algae in turbulent suspension. In large channels, however, it is difficult to maintain this flow across all channels with several bends so that a minimum velocity of 15 cm s^{-1} is recommended to achieve a speed of 5 cm s^{-1} in all locations of the pond. There is no evidence that the current velocity of the algal culture between 10 and 30 cm s^{-1} *per se* influences the yield of algae biomass, so that the mixing velocity is primarily determined by other considerations such as pond engineering, nutrient supply and addition of CO_2.

In unlined ponds, sediment suspension and transport may become a problem because of light reduction, channel erosion and contamination of the harvested algae biomass. Several experiments have been performed to predict the critical hydraulic parameters at which transport of soil and silt will begin. Particles become more easily suspended with size down to 0.1 mm diameter; at smaller sizes the particles tend to agglomerate. Flow rates exceeding 15 cm s^{-1} risk sediment transport so that the upper limit for mixing velocities in unlined ponds is about 20 cm s^{-1}. The power required to mix the medium increases as the cube of the velocity, while the incremental yield is linear with velocity. Hence, it is economically advisable to minimize the mixing velocity whenever possible in order to reduce the overall production cost. The costs for mixing are related to the power requirements which can be calculated on the basis of Δd and the volume of water to be agitated:

$$P = QW \, \Delta \, d/102 \, e,$$

in which P is the power required (kW), Q the water volume agitated (m^3 s^{-1}), W the specific weight of the water (about 999 kg m^{-3} at 15 °C), Δd the change in depth (m) before and after the location of the mixing device, 102 the conversion factor to convert m kg s^{-1} to kW, and e the efficiency factor of the agitator. For channel width w, depth d and flow velocity V, the quantity of flow Q is:

$$Q = wdV,$$

where w and d are given in meters (m) and V in m s^{-1}.

With a paddle-wheel efficiency of 0.5, a pond depth of 20 cm, a channel width of 6 m and a flow velocity of 15 cm s^{-1}, the power requirement would be:

$$P = \frac{w \times d \times V \times \Delta \, d \times W}{102 \times e} = \frac{6 \times 0.2 \times 0.15 \times 0.1 \times 999}{102 \times 0.5} = 0.35 \text{ kW}$$

or 8.4 kWh d^{-1} or 6.5 kWh ha^{-1} d^{-1}. If the mixing velocity is doubled, the energy required to agitate one hectare of culture is increased nearly ten-fold.

A somewhat simplified equation for the calculation of the power requirement for mixing is given by Borowitzka & Borowitzka (1989) as:

$$P = (9810 \, A \, V^3 n^2)/d^{0.3} \, e,$$

where P is the power (W), A the pond area (m²), d the depth of the culture, V the mixing velocity (m s⁻¹), n the Manning friction factor and e the overall mixing system efficiency, which is given as approximately 30–40% for paddle wheels. It can be seen that the power requirement varies as the cube of the mixing speed and increases slightly at lower depths.

Several different techniques are available to create the necessary turbulent flow in the algal pond. The specific choice will depend on the particular situation, i.e. location of the plant, type of pond, availability of energy, economical considerations or type of algal species cultivated.

The importance of the last factor may be illustrated by the difficulties in developing mixing requirements suitable for the cultivation of *Dunaliella*. The lack of a rigid cell wall of this alga, which results in cellular fragility and hence sensibility to mechanical forces, made mandatory the exclusive use of paddle wheels, which run at low speed maintaining a low linear liquid velocity of about 10 cm s⁻¹. Other mixing regimes such as air-lift pumps or centrifugal pumps resulted in severe damage of the algal cells. The conclusions to be drawn from the present state of experience are that turbulence in the culture created by mixing is essential and that increasing the vigor of the turbulence to a certain extent increases yield, but simultaneously increases the production costs. Hence, it can be stated that the issue of productivity enhancement by mixing is by no means settled and there is a general concensus that this subject deserves high priority for research. The major problem in this regard is the fact that the very intensive mixing required to achieve maximal benefits from intermittent illumination would require high mixing energies, complicate the design of the pond and increase the operating (power) costs. Nevertheless, the concept of producing organized mixing effects in the pond by, for instance, contouring the pond bottom or introducing baffles, deflectors or similar obstacles into the current of the agitated algal suspension, has to be seriously considered in future research.

Various agitation methods will now be described, some of which are commonly used in pilot and commercial algal plants whereas others are still in an experimental stage. The latter are employed particularly in developing countries to decrease algal production costs. They are shown to illustrate the many possibilities of creating turbulence in algal cultures and may serve as suggestions and stimulus for others working on the development of alternate mixing devices.

The common techniques are as follows.

1. Paddle wheel driven by an electrical motor. Although relatively expensive, this method is being used in most of the existing algal projects. The power demand is about 600 W for a 100 m² pond.

2. A combination of pumps and gravity flow, which has been adopted in some large-scale production units. The power demand varies between 100 and 200 W for a size of 100 m².

Aeration provided by pumps or aerators designed for mixing is of limited value in large ponds because of the small percentage of the pond area being mechanically mixed. The maximum head loss or pumping head in shallow ponds is limited by the ability to maintain sufficient flow to the mixing device, which controls the maximum channel length between mixing points.

3. Air-lift system. A power consumption of 195 W was estimated for a 85 m² pond based on a compressor efficiency of 70% and an air demand of 120 l s⁻¹.

4. Experimental devices.
   i) Free propeller with a power demand similar to that of the paddle wheel. Suitable for smaller ponds only.
   ii) Injector. With this method the medium is pressed through a nozzle, CO_2 is added in a separate chamber to achieve high turbulence and high CO_2 transfer rates. The jet introduces a flow in the pond. Corresponding to the high level of turbulence established the power demand of the system is high (1000–2000 W 100 m⁻²), which prohibits the wider application of this system.
   iii) Manual agitation with simple pumps or brooms. These techniques require practically no financial input but are of low efficiency and applicable for small cultivation areas only.
   iv) Paddle wheels or vertical stirrers driven by unconventional means such as savonius rotors, wind mills, solar cells, clock work, animals and man.

Paddle wheel

The most common devices for mixing algal cultures are paddle wheels, which were initially developed for the agitation of small experimental tanks and ponds. These paddle wheels consist of simple partial-depth blades or high-speed rotors. Because of the backflow around and under the blades, which prevents the creation of significant heads, the original type of paddle wheel is not suited for the agitation of larger ponds. The creation of more head by means of fast-rotating rotors is not practical for algal cultivation because it will cause damage to the algal cells owing to high shearing forces.

An improved paddle-wheel design was developed and successfully tested in larger cultivation ponds by Dodd (1986). The principle of this type of design is that of a positive-displacement vane pump, capable of developing the head needed to establish flow in a long pond. The improvement is that the ends of the paddle blades have a very small clearance with the walls and the curved-bottom section of the pond. Flexible blade-tip seals may be used to improve the performance of the device. The arc length is sufficient to prevent the water from escaping from between adjacent blades so that the speed of the blades has to be only slightly higher than the velocity of the medium.

For increased channel widths, special care has to be taken to prevent misalignment of the center shaft of the paddle wheel and wearing out of roller bearings. The shaft size should be chosen to control deflection at midspan, torsion, and bending imposed by the resistance of the water on the rotating blades. The blades should be bolted to hubs fixed on the main shaft in order to permit easy removal for cleaning, repair and tip-clearance adjustment.

For the drive of larger paddle wheels a double-chain-drive low-speed reduction through a jack shaft with sheer pin for overload protection is recommended. Variable-speed drive units are preferable because of the possibility to vary the mixing velocity. For a moderately large paddle wheel, which creates a head difference of more than 15 cm, Dodd (1986) has given some basic construction parameters. Eight paddles are used in the typical paddle wheel, displaced at mid-channel by 22.5° to reduce cyclic loads on the drive as the blades enter and leave the water. A mid-channel baffle prevents flow around the blades. The width can be estimated by the sum of the depth of the pond-bottom depression, water depth, lift and free-board allowance. The minimum tip radius will be about 1.0–1.2 m for an eight-blade paddle wheel. As construction material, fiberglass (preferably sheet fiberglass with corrugations) is recommended because of its weight, strength and durability. If this material is not available, simple plywood may be used; however, shorter duration and warping may cause problems.

Paddle-wheel efficiency, determined experimentally in a 0.25 acre pond, has been reported to range between 25 and 84%, being directly proportional to mixing velocity and pond depth. The overall power requirement for paddle wheels at a pond roughness coefficient of 0.025 has been determined at 20 kWh ha^{-1} d^{-1} at a mixing velocity of 15 cm s^{-1}, and 160 KWh ha^{-1} d^{-1} at a velocity of 30 cm s^{-1} (Benemann, 1986).

Air-lift systems

The air-lift system consists of a vertical conduit open at both ends, through which gas is sparged to move liquid through suitably arranged inlets and outlets.

The principle of the air-lift system is shown in Fig. 10.33. As can be seen, there is a down-flow part and an up-flow part in the lift area. The fractional void volume of gas in the rising dispersion must at a minimum provide enough buoyancy to overcome the pumping head. Maximum pumping efficiency can be expected when the liquid rise velocity is approximately equal to the flow velocity in the pond. Analysis of pump performances have demonstrated that the down-flow section should have the same cross-sectional area as the pond to ensure minimal energy losses in this part of the lift (Märkl & Mather, 1985). The up-draft tube height should be at least

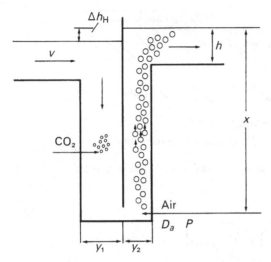

Fig. 10.33. Principle of the air-lift system.

four times, preferably more, the pumping head to be overcome. The greatest efficiency of this system will be at a mixing velocity of about 30 cm s^{-1}. Even at this mixing speed, efficiency will be only 50% and the power requirement will be almost twice as high as for paddle wheels.

If CO_2 is introduced into the down-flow, the velocity of the medium is equal to the ascending velocity of the CO_2 bubbles so that a very long residence time of the gas inside the channel will be achieved. In technical systems, bubbles with a diameter between 3 and 6 mm are produced that have an ascending velocity of about 23 cm s^{-1}. If the velocity of the ascending medium is kept in the same range, the gas bubbles will remain at the same location and become smaller until they disappear.

Calculations show that the amount of CO_2 washed out to the open air in the up-flow part of the system is only 5%. Assuming additional 10% losses in the down-flow section and another 5% at the open surface of the pond, the CO_2 utilization in this system is about 80%. The delivery head of the pump is calculated by adding the hydraulic head to the friction losses of the channel system. For a model pond, the optimum design parameters to achieve lowest power demand are $y^1 = 0.15$ m, $y^2 = 0.1$ m, $x = 0.8$ m. The air demand is $D_a = 121$ s^{-1} at a pressure of $P = 1.12$ bar. Although the system is simple and easy to operate, it has the disadvantage of being difficult to clean. The ditches require a special design of pond with additional excavations and foundation work that increase the construction cost of the pond.

The employment of the air-lift system was introduced by Clement & van Landeghem (1970) for the cultivation of *Spirulina* (Fig. 10.34). In this construction, the algal ponds have ditch-like depressions on both the front sides. Into these sections, CO_2-rich waste gas was blown, causing a

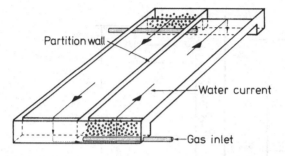

Fig. 10.34. Design of algal pond with air-lift mixing system.

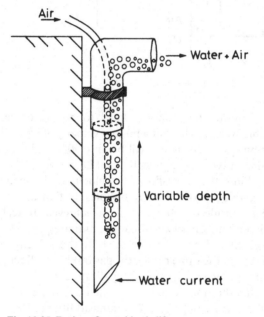

Fig. 10.35. Design of portable air-lift pump.

continuous circulation of the medium because of the water-lifting effect of the rising gas bubbles. The problem of providing additional impressions in the pond can be overcome by a modified portable design of the air-lift pump, which can be installed in normal ponds with even bottoms (Fig. 10.35). This type can be constructed from PVC-piping; the length of the pipes can be adjusted to the height of the water column by telescoping of the parts of the tube. Compressed air is led into the tube by small-diametered pores on both the tube and the hose; the position at the end of the hose determines the efficiency of the system. The flexibility of the device, its simplicity of construction, dismantling and adaptation to ponds of different sizes favours this system especially for medium-sized ponds, which are used commonly in aquaculture for the production of live food sources for

Fig. 10.36. Manual stirring of *Spirulina* culture with brooms.

larval molluscs and crustaceans. Besides agitating the medium, the system also effectively aerates the algal cultures and provides additional atmospheric CO_2.

Other systems

Agitation by manpower

In cases where continuous mixing of the algal medium is not required, manually operated agitation systems can be utilized. The simplest method is intermittant manual stirring of the culture by means of brooms or similar instruments for short periods during the course of the day and is suitable for small ponds used for the production of *Spirulina* (Fig. 10.36).

More advanced systems have been reported from Madras (India), where bicycle-powered agitation is provided for circular ponds (Fig. 10.37). The device consists of a wooden cantilever construction with one end swivelling in a pipe at the centre of the pond and the outer end being fixed to a bicycle and pulled around the pond. An array of paddle planks, attached to the rotating beam, agitates the medium.

Fig. 10.37. Pedal-powered agitation of algal suspension in a circular pond by means of a bicycle (MRCR, Madras, India).

Fig. 10.38. Mixing of algal suspension in twin ponds by means of a manually operated pump (MCRC, Madras, India).

At the same place another agitation technique was tested: the mixing of algae cultures in twin ponds by pumping the medium from one pond to the other one (Fig. 10.38). For this, two reciprocating pumps were fitted in such a way that they could be operated simultaneously with a common lever by two persons; the pumping head being very low, the pumps work with ease. Since the pumping site and overflow are located at the two opposite ends of the pond, it is ensured that the medium flows along the length of the individual channel.

Fig. 10.39. Mixing of algal suspension in twin pond by wind-operated pumping (MRCR, Madras, India).

Wind energy

Some of the manpowered agitation devices described above can also be driven by wind energy, preferably in areas with frequent winds. The operation of the two-man pump, for instance, can be performed by connecting the system to a battle-axe windmill as shown in Fig. 10.39.

Another wind-operated agitation device uses vertical savonius rotors. The very simple model shown in Fig. 10.40 is made out of used oil drums cut into two halves and held between two cross bars. The bottom of the vertical axis is ground suitably to form a pivot end that rests in a simple bearing, the top end being held by a wooden construction. At a moderate wind velocity (4 m s^{-1}) the stirrer is agitated at a speed of 30 rpm.

Clockwork

The prototype of a unique agitation system was constructed by Fox (1985) for the production of *Spirulina* as part of an integrated village health system. The driving force of this innovation is a mechanical clockwork device, made of bicycle and motorcycle gear wheels connected by bicycle chains, a heavy weight and a paddle wheel (Fig. 10.41). A weight of about

Fig. 10.40. Stirring of algal suspension in experimental ponds by means of savonius rotors (MRCR, Madras, India).

150 kg is hoisted on an iron mast up to a height of about 6 m, either by cranking the shaft of the transmission or by use of a separate system of pulleys. While the weight descends, it drives the paddle wheel for about 7 h. The water inertia acts as a control and the paddle speed is dependent on the depth to which the paddles are immersed into the water. With this construction the culture is kept moving at between 5 and 10 cm s^{-1}.

Solar cells
For the agitation of smaller volumes of algal suspension (inoculum ponds) another unconventional concept in algal cultivation was realized: the driving of paddle wheels by solar energy (Fig. 10.42). A photovoltaic unit which provides about 160 W was connected to a regular 12 V windshield wiper motor, which is powerful enough to turn the paddle wheel in a 24 m^2 pond. This type of motor, however, is suitable for shorter operational times

a

b

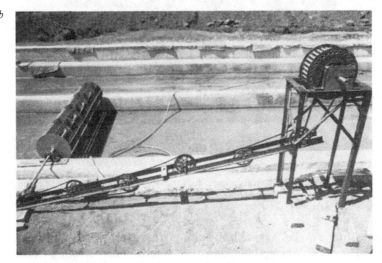

Fig. 10.41. View of mechanical clockwork device made of bicycle and motorcycle gears. *a*, General view of algal pond with mast for hoisting the weight; *b*, close-up view of reduction gear.

only. For permanent use, the installation of more durable motors is preferred. Simultaneously, these solar cell panels charge batteries that can serve as a power source during the night or during periods with unfavourable climatic conditions.

Based on the same concept, i.e. converting solar energy into electricity by photovoltaic elements, but different in its realization, is the mixing device that was developed recently in Germany especially for stirring ponds

Fig. 10.42. Photovoltaic unit which provides 12 V (about 160 W) electric power to a motor driving paddle wheels (Karla, India). (Photograph courtesy of Dr R. Fox.)

consisting of one channel only, although it can be used in raceway-type ponds as well. The system consists of a rectangular floating frame made from wood or/and joined plastic pipes of a width that corresponds to the width of the channel of the algal pond. The movement of the float is guided by castors, which are mounted at its outer corners and move freely on guide rails fixed at the inner wall of the channel. A panel of photovoltaic cells is mounted flat or at an inclined angle (corresponding to optimal light utilization at the respective latitude of the location) on top of the float. The system is driven by an electric motor, connected to an impeller mounted about 10 cm beneath the float. The rotation of the impeller depends on the direction of rotation of the motor which is controlled by two microswitches, installed at the bow and stern end of the float and which are operated by touching the corresponding end wall of the pond. By reversal of the rotation of the impeller, the float moves back and forth in the pond, the speed of which can be regulated by appropriate transmission. The stirring effect can be increased by mounting additional baffles under the float. The device, especially suitable for areas with long periods of sunshine, was trouble-free and provided sufficient turbulence in *Spirulina* cultures.

While the clockwork agitation device represents a system that can be adopted instantly even in rural areas, the utilization of solar panels opens up new possibilities. At present, the price of photovoltaic cells limits their application to experimental stations. It can be assumed, however, that in the near future these cells will become more durable and cheaper so that they can be employed at low cost even in remote areas.

It has to be stressed once more that most of these simple systems do not provide uniform and intensive mixing of the algal cultures, but are sufficient to ensure satisfactory growth, mainly for the cultivation of *Spirulina*. Comparative studies performed with this alga between cultures mixed continuously by paddle wheels and those stirred manually in intervals have shown that the latter yielded about 20% less algal biomass; but this reduction will be compensated for by savings in energy.

Mixing board
A means suitable for manual agitation of smaller rectangular ponds is the so-called drag board or mixing board. The device consists of a pull bar built as a triangular prism of wood, weighted so as to remain submerged with its base flat against the bottom of the pond. While moving the construction, a surface inclined at 45° digs the water up from the bottom and pushes it over the top of the bar. The construction is pulled the length of the pond by means of a pulley system inside the basin sharing one axel with a parallel pulley system situated outside.

A similar, motor-driven concept is being used in larger ponds in Brazil and Chile. For the mixing of the algal system, a drag board was developed which closes the pond cross section except for a slit of a few centimeters above the bottom. Slowly moved, the board forces all of the water to run through this slit, thereby creating a turbulent backwhirl (Fig. 10.43). According to the published data, the energy requirement is only 20% of the energy needed for comparable agitation with paddle wheels. This figure is extremely low and it remains to be proven whether this can be established in long-term experiments.

Another alternative is the utilization of large, low-head centrifugal pumps, as practiced with sloped ponds. This system requires relatively deep sump pits (1–2 m) as well as reduction of the channel width at the pump. Further disadvantages are that the mixing speed cannot be easily controlled and that the centrifugal pumps might break fragile cells like *Dunaliella*.

It has already been mentioned that the photosynthetic productivity of algal cultures can be increased by providing the light in the form of short flashes. A simple device for producing such a flashing (better called

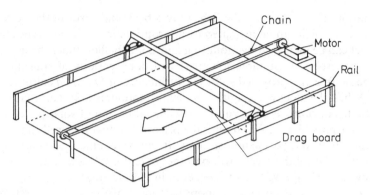

Fig. 10.43. Design of algal pond with drag-board agitation system.

'intermittent') light effect by creating a non-random vertical mixing in outdoor algal cultures was described by Laws *et al.* (1983). The construction consists of an array of foils similar in design to aeroplane wings, which are placed across the current of the algal suspension at a distance of 1.2 m (Fig. 10.44). The length of each foil and the gap between the foils are equal to the culture depth so that circular vortices created effectively mix the suspension from top to bottom. The uniform mixing results from vortices with the rotational axis parallel to the flow direction that are produced because of the pressure differences created as water flows under and over the foils. At a current velocity of 30 cm s^{-1}, the rotation rate of the vortices are reported to vary between 30 and 60 cycles s^{-1}. According to the authors, the production of *Phaeodactylum tricornutum* could be almost doubled by employing this system.

The introduction of such foils or similar devices seems to be an economical means of inducing non-random mixing, irrespective of the method used to create flow of the culture, if their benefits in terms of costs and actual effectiveness are substantiated. The problem of productivity enhancement by mixing is not settled yet. The major problem in this regard is that the very intense mixing required to achieve maximal benefits from intermittent illumination would require high mixing energy and a more complicated design of the pond, which would increase the algae production costs. The conclusions to be drawn from the present state of knowledge are that low mixing velocities and small turbulences increase productivity over laminar systems, that turbulent mixing and increasing the vigor of turbulence may improve the productivity, and that organized mixing for instance by the introduction of special devices may have some practical benefits but that no practical means of achieving this has yet been demonstrated.

10.4.   CO$_2$-inlet systems

The question of how to supply CO$_2$ to algal cultures is a key engineering consideration. Three major factors must be considered when discussing

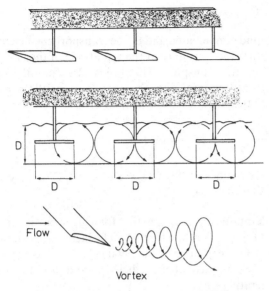

Fig. 10.44. Design and working principle of foils introduced into algal raceway ponds in order to produce an intermittent light effect by the formation of a vortex. (After Laws *et al.*, 1983.)

CO_2 transfer from air into the pond: the mixing regime, the CO_2 concentration in the pond (dependent on pH and alkalinity), and the effect caused by the reaction of dissolved CO_2 with OH^- to produce bicarbonate.

Several techniques have been developed to distribute the carbon dioxide into the culture medium. Principally, two different attempts can be distinguished to achieve an efficient utilization of the carbon dioxide by the algae:

a) active gas transfer through sparging small gas bubbles into the medium or spraying the liquid through the gas phase; and

b) passive transfer by creation of large contact areas between a CO_2 atmosphere and the surface of the culture medium.

Before describing some of the most common carbon-dioxide distribution systems, a few theoretical reflections should be made first.

The overall balance of carbonates in algal cultures is determined by the influx through the supply system, the carbonate fixed through photosynthesis by the algae as organic carbon, and the losses by desorption through the interface between the algal suspension and the air. The fixation of carbonate via photosynthesis depends upon the growing conditions of the algae (illumination, pH, temperature, nutrients, etc.) and is expressed by the increase of algal biomass. If carbon dioxide is added as pure gas into the medium, the transport of CO_2 from the gas phase to the algal cell can be expressed by the empirical formula according to Märkl & Mather (1985):

$$\tilde{n}_{CO_2} = \beta_L \times A(c_{CO_2 \text{ bubble}} - C_{CO_2 \text{ cell}})$$

where A is the gas–liquid contact area and β_L the transport coefficient. The subscript $_L$ indicates that the resistance at the liquid side of the gas–liquid interface is controlling the transport. This holds also for other gases, namely oxygen or nitrogen, so that β_L has the same value for all these gases. The concentrations at the gas side can be defined by:

$$c_{CO_2 \text{ bubble}} = \frac{P_{CO_2 \text{ bubble}}}{H} \times 55.56 \text{ mol l}^{-1}$$

where　H is the Henry coefficient (for $CO_2 = 1.64 \times 10^3$, $O_2 = 4.38 \times 10^4$ and $N_2 = 8.65 \times 10^4$ bar).

From these considerations it can be seen that for a given partial pressure the mass transfer for O_2 is smaller than for CO_2. The factor $c_{CO_2 \text{ cell}}$ indicates the concentration of CO_2 very close to the algal cell. It is this value that actually determines the photosynthetic efficiency and not the partial pressure p_{CO_2} in the aerating gas.

It has been calculated that a concentration of CO_2 as low as 10^{-6} mol 1^{-1} (30 p.p.m.) is sufficient to maintain unlimited photosynthesis of the algae. On the other hand, experiments with different O_2 concentrations in the medium have shown that the photosynthetic efficiency is increased by 14% if almost no O_2 is present in the medium but is reduced by about 35% when the medium is saturated with 100% O_2. In outdoor systems, the aim should be to keep the O_2 concentration in the culture no higher than the saturation concentration with air (21%). With larger cultivation units it is almost unavoidable that the CO_2 concentration is not uniform at all locations of the pond. Assuming that the CO_2 is introduced into the pond at one point, a certain concentration will be established shortly after this point. As the medium moves around the pond it is losing CO_2 because of fixation by the algae and loss to the atmosphere (Fig. 10.45). In a well-balanced system the CO_2 concentration in the medium before reaching the aeration point should not be lower than the critical CO_2 concentration needed for unlimited photosynthesis. If, on the other hand, the CO_2 concentration is kept at excessively high concentrations, too great a loss to the air will occur. The correlation between initial CO_2 concentration, distance between aeration areas and CO_2-fixation by algae is shown in Fig. 10.45 for an experimental pond (culture depth = 15 cm, mixing velocity = 30 cm s^{-1}) with an assumed algal growth of 30 g m^{-2} d^{-1}. It can be seen that with an initial CO_2 concentration of about 40×10^{-6} mol 1^{-1} the critical minimum concentration is reached after about 45 m. For simplification, the transformation of CO_2 to carbonate is neglected, which is correct at pH values lower than 5.3. At pH 6.3 the carbonate pool is equivalent to the CO_2 pool. In the case of higher pH values, CO_2 is stored almost completely in the form of carbonate.

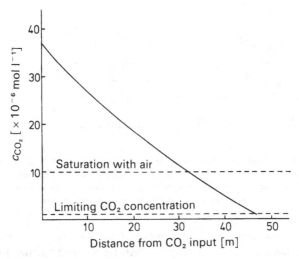

Fig. 10.45. CO_2 concentration at different locations in a pond in relation to the distance from CO_2-inlet point.

CO_2-aeration systems

The simplest way to introduce a gas into a liquid is by bubbles. In laboratory algae cultures, it is common to aerate the medium either with pure CO_2 or with mixtures of CO_2 and compressed air. During the initial stages of outdoor algal cultivation, the same concepts for aeration were employed as were used for indoor cultures. Therefore, sintered porous stones or pipes, commonly used in aquariums, were adapted for the distribution of fine gas bubbles into the algal ponds (Fig. 10.46A). A similar method with a better efficiency can be used in connection with air-lift agitation systems. If the CO_2 is injected into the downward flow of the suspension at the pump section, the retention time of the gas in the culture can be increased (see also the section on agitation, p. 95). Unfortunately, all these methods are afflicted with several disadvantages. In shallow ponds they result in high losses of CO_2 to the atmosphere because of the short retention time of the gas bubbles in the algal suspension that does not permit complete absorption of the CO_2. It was found that under working conditions it was difficult to achieve a utilization of more than 10% of the CO_2 supplied. In addition, there are several technical problems in mounting pipes, tubes etc. at the bottom of the pond, and the porous material used to distribute the gas into the medium tends to get clogged by the algae and other debris so that permanent cleaning is required.

A modification of this gas inlet system was the introduction of long PVC pipes (connected at one end to a CO_2 line and sealed at the other end) that were placed longitudinally at the bottom of the pond. A line of fine holes was drilled into the upper part of the pipes so that the gas could be distributed over longer parts of the pond ensuring a better efficiency of CO_2

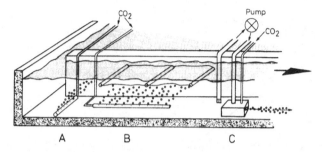

Fig. 10.46. Different systems for CO_2 aeration in algal ponds. A, Sintered stone; B, porous pipe under submerged plastic sheets which traps the CO_2 bubbles and increases contact time between gas and water; C, pressure pump which injects air/CO_2 mixture at high speed thus achieving areation and mixing.

utilization. However, substantial amounts of CO_2 are lost to the atmosphere with this method also. These losses can be reduced further by the installation of a covered sparging–diffusion system, consisting of a transparent plastic sheet tightened to a frame that would be mounted about 5 cm below the waterline above the pipes along the pond (Fig. 10.46B). The small gas bubbles, injected through the pipe into the culture, are trapped under these sheets in the form of larger gas bubbles, which slowly move along with the current of the culture. This effect increases the contact time between the gaseous CO_2 and the suspension, resulting in a more efficient utilization of the CO_2. To maximize transfer, additional liquid turbulance would be created under or next to the cover. To minimize oxygen release under the cover, an area of turbulence may also be appropriate some distance upstream to facilitate O_2 release.

No exact figure can be given on the total area to be covered by such a transfer system to achieve maximum CO_2 utilization because only scattered experimental data are available that vary within one order of magnitude. It is expected that CO_2 transfer can be accomplished within 0.5–5% of the total surface area. Despite these installations, considerable amounts of CO_2 are still lost at the end of the plastic sheet where the unabsorbed gas escapes into the atmosphere.

With the latter method, it is important that the plastic sheet is submerged and fixed below the surface of the algal suspension. The submerged sheet, which should cover the full width of the pond channel, ensures that the CO_2 bubbles do not leak out but remain in contact with the medium for an extended period of time. Furthermore, if parts of the sheet rise above the water surface, a 'greenhouse effect' will occur. The medium will get heated up so that it evaporates and subsequently condenses on the bottom side of the sheet in the form of small water droplets that will act as a mirror and reflect the light, thus reducing the radiation intensity reaching the algal culture.

A further possibility for supplying algal cultures with CO_2 is by combining agitation and aeration. It has already been mentioned that algal cultures can be agitated by employing different methods of pumping. One of these systems is a kind of injector in which the algal suspension is pumped with pressure through a nozzle. With this method, CO_2 can be added in a special chamber where high turbulence and good CO_2 transfer rates can be established (Fig. 10.46C).

To achieve better growth rates two types of injectors should be used simultaneously in one pond system. With one injector, CO_2 is introduced into the culture whereas in the other air is sucked into the turbulence chamber. The latter type produces an increased gas–liquid contact area for better O_2 exchange, which helps to avoid oversaturation of the culture with photosynthetically produced oxygen.

For larger cultivation areas, sloped ponds with gravity flow have been developed where the suspension is pumped through pipes from the lower part of the pond to the upper part. If CO_2 is injected at the lower end of the pipe, the gas will be trapped and flow together with the suspension through the full length of the pipe, resulting in an extended contact time between algae and CO_2. The bubbling method can be improved further if the algal suspension is enriched with CO_2 in a closed column. Such gas absorbers, installed in a by-pass of the recycling system, were first used in Czechoslovakia (Simmer 1969).

By injecting the CO_2 through porous tubes into the pond and installing floating plastic sheets beneath the surface the amount of gas needed to obtain satisfactory growth rates is in the range of 10–20 ml h^{-1} l^{-1} of culture medium. By supplying CO_2 continuously for 8 h per day from 9 a.m. to 5 p.m., more gas is added than can be utilized during this time. Because the pH of the medium is proportional to the amount of dissolved CO_2, monitoring devices can be used to maintain the pH at a constant desired value by the injection of controlled amounts of CO_2 into the medium. If the gas flow rate is kept constant and a recorder connected to a solenoid gas valve that records the time of day when and for how long the gas is injected into the culture, the amount of gas supplied from morning to evening can be calculated.

Experiments showed that the number of times when the valve has to be opened to maintain a pH of 7.6 increased during the course of the day and reached its maximum in the early afternoon hours between noon and 4 p.m. It was observed that in cultures cultivated for 5 days under the above conditions, CO_2 was added for a total period of only 4 h per day. No differences in growth could be found between these cultures and cultures supplied with CO_2 continuously for 8 h per day. As it is too costly to equip a number of ponds with such a pH monitoring device, a simple method that helps to minimize the waste of CO_2 is recommended. Based on the above experience that CO_2 supply for 4 h per day is sufficient to maintain good algal growth, the gas can be added in hourly intervals to the culture, i.e. 9 a.m. to 10 a.m., 11 a.m. to 12 a.m. . . . etc. As can be seen from Fig. 10.47, no

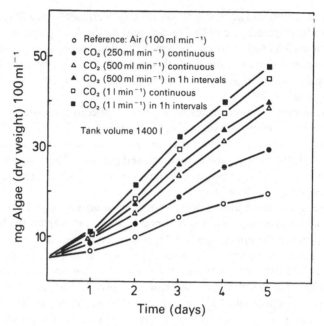

Fig. 10.47. Effect of various CO_2 concentrations and supply regimes on the growth of *Scenedesmus*.

substantial differences were found between continuous supply and this procedure, which saved 50% of the CO_2. Of course, this procedure can be improved by increasing the number of the shorter periods of gas supply, which can be adapted better to the peak hours of photosynthesis and CO_2 requirement, determined by the particular environmental conditions and culture regimes.

Summarizing the efficiencies of all techniques used at present to supply CO_2 to algal cultures, it can be stated that their performance is still not satisfactory because they all fail to adjust exactly the amount of CO_2 supplied to the actual requirement of the growing algae.

More efficient is the second type of aeration system, namely enlarged contact areas between CO_2 atmosphere and culture medium, based on the principle of a bell-gasometer, consisting of an open vessel, introduced upside down into the algal suspension and filled with CO_2. With this construction, the gas enclosed below the container is transferred into the liquid surface by direct diffusion.

A floating CO_2 injector, which fulfils all the requirements for a uniform distribution of CO_2 over large pond areas with minimum losses, was developed in Peru and described in detail by Vasquez & Heussler (1985).

At present, this construction seems to be the most efficient means for supplying CO_2 to shallow open ponds. Moreover, the construction of the gas exchanger is simple and can be made with material available locally,

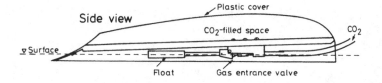

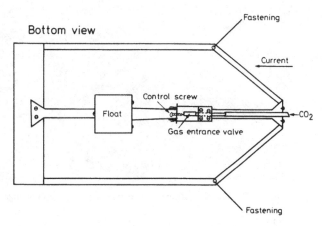

Fig. 10.48. Schematic design of floating CO_2 injector.

even in developing countries. The design of the device is shown schematically in Fig. 10.48. It consists of an inverted (bottom open) compartment, preferably made from sealed PVC pipes and covered with transparent sheeting, which floats on the surface because of the development of a gas cushion under the cover. The compartment is filled with pure CO_2 by first submerging the complete system (allowing the air out through a gas valve) and then filling it through a ball-valve system. The valve, for instance a standard ball valve obtainable from a plumber, is mounted in such a way that when the float drops, i.e. when the device emerges due to its buoyancy, the valve is shut and vice versa. The system can be adjusted (vertically and horizontally) to determine the floating height (maximum gas volume) and to regulate the counter pressure of the CO_2 in the supply hose. Spoilers across the injector produce a high turbulence in the running algal suspension that is necessary for efficient gas transfer into the liquid. Because of the shape of the rear end of the injector the turbulence of the leaving algal suspension is slowed down to avoid excessive desorption losses to the atmosphere. The installation of floating injectors in the sloping algal pond in Peru is shown in Fig. 10.49. As can be seen, the surface area covered by the floats is negligible and does not affect the photosynthetic efficiency of the system due to possible shadowing.

The flow of carbon dioxide through the gas–liquid interface, either the influx from the floating injectors or the desorption to the open atmosphere, can be described by the equation:

Fig. 10.49. View of floating injectors installed at the sloping algal pond in Peru.

$$Q = kA(C_s - C_d)$$

where Q = mass flux of carbon dioxide (mM l^{-1})
k = mass transfer coefficient (M min^{-1})
A = interface area (m^2)
C_s = saturation concentration of dissolved CO_2 in the liquid in equilibrium with the partial pressure of CO_2 in the gas phase (mM l^{-1})
C_d = instantaneous concentration of dissolved CO_2 in the liquid (mM l^{-1}).

This flow is directed towards the lower CO_2 concentration. Under the injectors the partial pressure of CO_2 must be high to attain a high C_s and a satisfactory efficiency of gas transfer per area unit. On the other hand, C_d should be low to avoid excessive losses of CO_2 through the large pond surface in contact with the atmosphere with a low partial pressure. If the CO_2 flow into the injector and the partial presure in the gas mixture enclosed are known, the mass transfer coefficient for a given injector area can be estimated as:

$$k_i = Qt / [A_i(C_{si} - C_d) \times 1000] \text{ (M min}^{-1})$$

where Qt = CO_2 inflow (mM min^{-1})
A_i = area of injector (m^2)
k_i = mass transfer coefficient of injector (M min^{-1})
C_{si} = saturation concentration of dissolved CO_2 in equilibrium with the partial pressure below the injector (mM l^{-1})

C_d = real concentration of dissolved CO_2 in the algal suspension ($mM\ l^{-1}$).

According to this equation the efficiency of the injectors depends on the mass transfer coefficient and the difference between the saturation concentration and the actual concentration of dissolved CO_2 in the suspension. For an optimum efficiency the difference $C_{si} - C_d$ should be as high as possible, which means that a high concentration of CO_2 has to be maintained in the gas phase below the injector. Assuming daytime temperatures of the culture medium between 25 and 30 °C and saturation concentrations of the injector between 15 and 25 $mM\ l^{-1}$ at the maximum concentration of 1.5 $mM\ l^{-1}$ of dissolved CO_2, flow rates from 70 to 130 mM of $CO_2\ min^{-1}\ m^{-2}$ of injector area can be expected. That means that under practical conditions 1 m² of injector area can supply 1.8–3.4 kg of CO_2 during 10 h to the pond. The losses of CO_2 from the liquid surface to the atmosphere are influenced by the concentration of dissolved CO_2. At the temperatures given above, the recommended maximum concentration of 1.5 $mM\ CO_2\ l^{-1}$ will cause losses around 7 $mM\ min^{-1}\ m^{-2}$.

For further assessment of CO_2 utilization, the concentrations of dissolved CO_2, carbonate and the pH were monitored during the daytime in a dense, intensely growing *Scenedesmus obliquus* culture at the bow and the rear end of one injector of 1 m², floating on a sloping pond with a length of 100 m. During maximum illumination, the level of total carbonate was reduced by 0.2 $mM\ l^{-1}$ after this distance and the concentration of dissolved CO_2 by about 50%; the pH value increased by 0.3 units. It could be shown that 4 m² of injector area are required per 100 m² of pond surface to maintain a minimum CO_2 concentration of 0.2 $mM\ l^{-1}$ in an actively growing culture, even during the hours of maximum photosynthetic activity. A considerable part of this injector area is required to compensate the desorption losses; to meet the demand of an algal culture with a growth rate of 25 $g\ d^{-1}\ m^{-2}$, 1 m² of injector area could supply 40–75 m² of pond area.

If the CO_2 supply was switched off during the night, the total carbonate concentration increased and reached a concentration of about 2.0 $mM\ l^{-1}$ and the pH approached a value of 8.2. It was observed that the total carbonate level in the morning was higher than the concentration in the evening, presumely because of a rise during the first hours after sunset. This rise coincides with the time of cell division in the algal culture, and it was supposed that during this period the respiratory activity of the algae is enhanced, thus increasing the carbonate pool.

The initial design of the floating injector had one drawback in so far as the CO_2 in the compartment is continuously diluted with other gases, especially photosynthetically produced O_2 and nitrogen from the atmosphere. Fig. 10.50 shows the change of the gas composition in the injector (volume 44 l) as a function of operation time. It can be seen that the CO_2

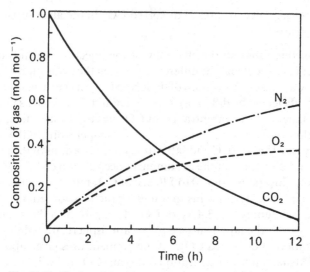

Fig. 10.50. Change of the gas composition in the floating gas exchanger (volume 44 l) as a function of operating time.

partial pressure decreases with time while the amounts of O_2 and N_2 increase. For sufficient supply of the algae, a CO_2 pressure of at least 0.5 bar is required. This minimum is reached after approximately 3–4 h. The efficiency of the system can be improved by increasing the volume of the injector (Fig. 10.51) in order to maintain sufficient CO_2 supply over longer periods of time. A still better performance, however, can be achieved by continuous de-aeration through a small hole in the cover of the injector. Depending on this leakage rate an equilibrium will be reached after a certain period of time (Fig. 10.52). The loss of CO_2 is relatively small, being about 4% of the total CO_2 supply.

Carbonation column

Another method for increasing the transfer of CO_2 through a neutral gas–liquid interface is by providing the lengthy time and wide surface area required to maximize transfer. Packed-column carbonation, which is often used in industry, is not suitable because its low pH frequently harms and damages the algal cells. An effective compatible method for transferring CO_2 to algal cultures is counter-current carbonation by means of carbon absorption columns. With this system the gas is injected as minute bubbles into a column of algal suspension; the velocity of the medium is adjusted in such a way that the bubbles of gas rising against the current remain in the water until fully absorbed (Fig. 10.53). The system is easy to build and to operate and permits the coverage of a large range of conditions. One of the main difficulties is the efficiency of the gas transfer in the column, which is

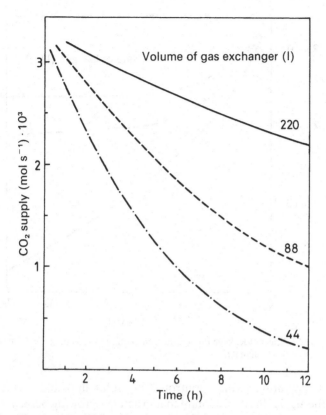

Fig. 10.51. Equilibria of floating gas exchangers in relation to their volume.

usually not very high, leading to considerable waste of compression energy and CO_2. It has been reported that by regulating the CO_2 input according to light intensity, a 30% reduction in expenditure could be obtained. In large plants several CO_2 injection points have to be installed; however, this is costly in terms of gas-distribution equipment.

10.5.   Effect of pH

The pH of the culture medium is one of the important factors in algal cultivation. It determines the solubility of carbon dioxide and minerals in the medium and directly and indirectly influences the metabolism of the algae. Algae exhibit a clear dependency on the pH of the growth medium and different species vary greatly in their response to the pH. *Cyanidium*, for instance, has its growth optimum at pH 2.0, whereas *Spirulina* grows well at pH values between 9 and 11.

The pH of algal cultures can be influenced by various factors such as composition and buffering capacity of the medium, amount of CO_2

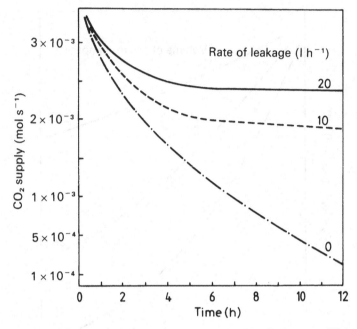

Fig. 10.52. Effect of leakage rate on the time required to obtain gas equilibrium in floating gas exchanger.

dissolved, temperature (which controls CO_2 solubility) and metabolic activity of the algal cells. It was found that in actively growing *Scenedesmus* cultures without addition of CO_2, the pH increased during the day to values up to 10, mainly because of the depletion of the anions NO_3^- and of CO_2 from the medium and the excretion of OH^- ions.

In *Spirulina* cultures, which are neither aerated with air nor with CO_2, the pH remains almost stable at 9.5. Any shift in the pH above 8.5 in the medium for *Scenedesmus* or *Chlorella* can be adjusted with suitable addition of phosphoric acid, hydrochloric acid or merely CO_2 supply. Adjustments of the pH in *Spirulina* cultures, if at all needed, which have a pH optimum of 8.5–10.5 and thrive at a high concentration, can be made with the addition of NaOH solution or with sodium bicarbonate (0.2 M).

A rise of the pH in *Spirulina* cultures above 10.5 does not necessarily mean that the photosynthetic activity of the algal culture has decreased to zero. At such high pH values CO_2 from the air is absorbed in sufficient amounts to maintain photosynthesis and productivity of about 7 g d^{-1} m^{-2}.

Dunaliella has a wide pH optimum between 7 and 9, but even at values below and above these figures good growth is maintained.

One of the common effects observed in culture media (especially those

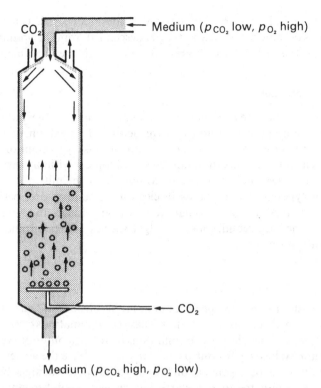

Fig. 10.53. Schematic diagram of a carbon absorption column.

with Ca^{2+} concentration > 10 mM or prepared from sea water) having pH values above 9, is the precipitation of several calcium salts, i.e. carbonates, phosphates and sulfates, leading to nutrient deficiencies and growth retardations or even algal flocculation, induced by the precipitating minerals. It is recommended to maintain a pH of 8 if the Ca^{2+} concentration exceeds 1 mM. Addition of mineral acids to media of high alkalinity needs the re-adjustment of the alkalinity to its original level by adding carbonates in stoichiometric amounts.

A more sophisticated method for maintaining the pH at a desired value is the introduction of pH controllers (pH states), which measure the actual pH of the medium and, as soon as a preselected pH is exceeded, open a solenoid valve that releases CO_2 to the medium until the desired pH is reached (see section on CO_2, p. 115). A modification of this method is the addition of HCl through such pH controller when HCO_3^- serves as the major carbon source. On the other hand, it should be kept in mind that high pH and alkalinities prevent night-time losses of respiratory CO_2 and always maintain considerable reserves of this nutrient.

In the case of toxic cyanobacteria, as for instance *Microcystis*, it has been reported that the pH influences both the growth rate and the toxicity in such a way that cells became more toxic as the pH descended from the optimum, suggesting that the slower the cells grew, the more toxic they became.

10.6.   Contamination

Certain kinds of contamination in outdoor algal cultures are inevitable in view of the non-aseptic conditions, where neither the medium nor the environment are sterile. In practice, it is therefore necessary to monitor and to control such contaminants to obtain an algal biomass without harmful impurities and to keep the contaminants within tolerable limits.

The major types of contaminants in clean algal cultures are, besides bacteria, other algal forms, zooplankton, viruses, fungi and insects, depending on the local conditions, the algal species cultivated and the particular cultivation system.

Unwanted algae

Invasion of outdoor ponds by species of undesirable algae is still one of the major problems in algae mass cultivation. These contaminations cannot be avoided completely since the growth conditions in an open outdoor system are suitable not exclusively for one particular species. Since the achievable algal productivity will be dependent on the particular species of algae being cultivated, the ability to grow a particular species is of fundamental importance.

Contaminating undesired algae impair the uniformity of the product and may reduce the yield but very seldom render the algal biomass useless; in extreme cases an alien algae species may become the dominant species and overgrow the original culture. Thus species control, i.e. the mainten-ance of systems of essentially unialgal cultures of the desired species, is one of the paramount problems in algal mass cultivation.

Information on such contaminations is very scarce, mainly because investigations about this aspect demand on cultures becoming infected, which occurs erratically. In cultures of *Scenedesmus* sp. the common contaminants are *Chlorella* sp., *Selenastrum* sp., and some species of diatoms. It has been reported from Japan that the maintenance of pure cultures of *Chlorella* requires frequent start-up of the culture with uncon-taminated inoculum. Benthic algae may occur on paddle wheels and on the walls of the algal ponds in the photic zone. The occurrence of other green algae in cultures of *Chlorella* or *Scenedesmus* normally does not represent a serious problem because none of the common algal contaminants is considered as harmful. The only exception would be contamination with certain toxic cyanobacteria such as forms of *Microcystis* or *Anabaena*. Fortunately, such a case has not been reported yet.

Spirulina cultures have become contaminated with *Oscillatoria* or with green algae at suboptimal growth conditions, for example bicarbonate concentrations below $15 \, \mathrm{g} \, \mathrm{l}^{-1}$.

Although it is impossible to prevent the contamination with undesirable algae, there are some precautions to be observed that help to keep such infections at an acceptable level. Plastic covers may delay or minimize the rate of contaminant invasion, but for practical reasons they are limited to small ponds and short periods of time only.

Better strategies are high concentrations of the inoculum, periodic cleaning of the ponds and creating specific environmental conditions that favor the growth of the desired algal species. This can be achieved through the absence of nitrogen fertilizer in cultures of nitrogen-fixing cyano-bacteria, high salt concentrations in cultures of *Dunaliella* and *Spirulina*, pH, or the nutrient supply. For instance the differences in affinities and short-term rates of uptake for ammonia of desired and undesired algae can be used as a mechanism to favor the growth of the desired organisms by selected regimes of ammonia and other nutrients supplied. By adjustment and controlled changes of the variables mentioned above, the dominance of the selected species may be favored and it also may be possible to counteract some of the deleterious effects on species dominance of uncontrollable variables such as temperature and incident sunlight.

In general, it seems that within certain limits, healthy algae establish monocultures that suppress the spreading of other species, perhaps by the excretion of growth-suppressing substances, although in almost all algal projects a succession of different species, particularly on a seasonal basis, has been observed. The contamination of *Spirulina* sp. with other algae is limited to very few forms, mainly because of the very specific culture conditions of *Spirulina* (high salinity and pH), which are not favorable for most other algae.

Mould, yeast and fungi

Contamination with yeast and mould in algae cultures, as determined by counts on dextrose agar, can be detected frequently in algae ponds; however, they do not present serious problems either for the performance of the microalgae or for the consumer. Among the moulds found, the species *Geotrichum candidum* in general grows on soiled parts of machinery used in food processing and is known to contaminate, for example, dairy products. Because of biochemical similarities between such products and algae – for instance, high protein concentrations – counts of *G. candidum* can be used as a parameter for the hygienic status of the algae plant.

Among the various contaminations encountered in large-scale algae cultivation, several cases of infection with parasitic fungi and zooflagellates have been reported. The fungi, commonly called chytrides, resemble a group of primitive fungi that have been detected on several algal species and often occur as epidemics, which sometimes culminate in the complete loss of cultures. The fungus most dangerous for cultures of chlorophyceae is *Chytridium* sp., which often appears together with the zooflagellate *Aphelidium* sp. Infections of *Scenedesmus* cultures with these organisms, have been detected in Germany, Thailand, Israel, South Africa and Peru. It seems that infections with *Aphelidium* sp. and *Chytridium* sp. are limited to cultures of chlorophyceae only; no reports could be found in the literature on similar contamination of *Spirulina* cultures. Because zoospores and zoosporangium of these contaminants are very small, it is quite likely that they have been overlooked in many other instances.

To illustrate the different stages of this parasitic infections, the life cycles of *Chytridium* sp. and *Aphelidium* sp. are shown schematically in Fig. 10.54.

The zoospores of *Aphelidium* sp. adhere to the algal cell wall, in the case of *Scenedesmus* sp. preferably close to the tips of the elongated cells. The algal cell wall is perforated locally and the protoplast of the parasite enters the algae through the orifice, leaving behind the fragile globular cell envelope of the zoospore. The *Aphelidium* protoplast grows inside the alga, exhausting almost all its content. Inside the algal envelope several new zoospores are formed and released through the infection canal. The early forms of an infection with *Chytridium* show a similar pattern as described for *Aphelidium*; in the later zoospore developing stage, however, large zoosporangia are formed at the outer surface of the infected algal cell from where the new zoospores are released. As the major part of the development of *Aphelidium* takes place within the algal cells and infected algal cells scarcely differ from healthy ones when seen under the light microscope, a staining method was developed to detect infections with *Aphelidium* (Merino *et al.*, 1985). The protoplast of *Aphelidium* (in the medium and inside the algal cell), fungi and bacteria can be stained with methylene blue (2%), showing a dense blue colour. Empty algal cells and some mucilaginous matter of other microorganisms, often found in debris of infected algal cultures, show a rose color. Healthy algal cells do not stain. Algal cells that died from high concentrations of fungicides (see below) stain rose-violet and can be clearly distinguished from infected cells.

Together with the advance of such an infection, heavy flocculation of the algal suspension occurs and the color of the medium turns brownish. Infected cultures show a marked decrease in oxygen evolution, even during the peak of photosynthesis. If these indications are observed it is already too late to combat the parasite; infections should be recognized as early as possible by daily microscopic examinations of the culture. The only biological control of an *Aphelidium* infection in an early stage is a manifold dilution of the culture with fresh medium. During exponential growth, the algal population multiplies at a higher rate than the parasite so that the infection might be overcome by repeated dilutions. As soon as the algal

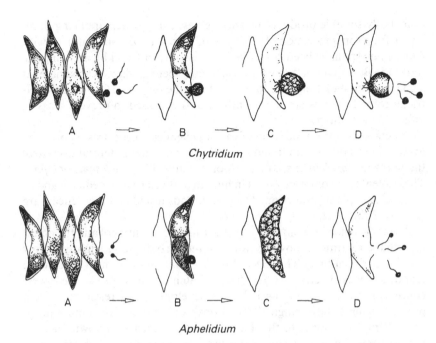

Fig. 10.54. Life cycle of fungal parasites *Chytridium* sp. and *Aphelidium* sp. in *Scenedesmus* sp.

Chytridium: A, zoospore attached on algal cell wall; B, growing *Chytridium* with haustorium, lysing algal cytoplasm; C, mature *Chytridium* zoosporangium; D, release of zoospores.

Aphelidium: A, zoospore attached on algal cell wall; B, Amoeboid *Aphelidium* protoplast phagocytizes algal cytoplasm; C, mature *Aphelidium*, mitosis; D, release of zoospores.

growth rate is lower than that of the parasite, the infection becomes troublesome. Hence it was concluded that the culture density must be maintained in the optimum (linear) growth phase by daily harvest. For treatment of an infected culture, further reduction of the population density is necessary until logarithmic growth is obtained. This method should also be applied when unfavourable environmental conditions prevail (shortage of sunlight, low temperatures, etc.). In general, culture density should not exceed 70% of the maximum algal concentration attainable under normal growth conditions. This method, however, is not very feasible in commercial algal production where dense algal cultures and high yields are required. The only remaining possibility for overcoming these problems would be the treatment of infected cultures with chemical agents. However, it has to be stressed that this method involves several problems because of the risk of residues in the final product.

At present, there are no chemicals available that are produced specifically to control infections of algal cultures. However, there are several potentially suitable products among the pesticides, which are being developed for crop protection. Infections of *Scenedesmus* sp. cultures with *Chytridium* sp. have been controlled successfully in Israel by applying the fungicide Benomyl (methyl 1-butylcarbomoyl benzimidazolecarbamate) at a concentration of 1 mg l^{-1}. In other places the fungicide Orthocid 50 (N-trichloromethyl-thio-tetrahydrophthalamid) was used to prevent massive infections with fungi.

A detailed study on the effect of various chemical agents was reported by Merino *et al.* (1985), who tested 15 pesticides and their potential to control the parasite *Aphelidium* sp. in outdoor cultures of *Scenedesmus obliquus*. These agents are commonly used in human and veterinary medicine and as fungicides in plant protection. The compounds tested and their effects are listed in Table 10.2.

The most efficient chemicals were the fungicides Antracol and the liquid Karathane (Karathan powder causes some turbidity and slightly yellow staining of the culture). The effect of Antracol lasted for about 4 days, while Karathane was effective for 5–6 days in batch cultures even at the lowest concentration tested (0.7 mg l^{-1}). No development of resistance of the parasite against these fungicides was observed. The active compound of both chemicals belongs to the thiocarbamates, which are known for their low toxicity and have been widely used in agriculture for many years. However, it has been reported that one of their metabolites is ethylene-thiourea, which is suspected to be carcinogenic under certain circumstances.

The decomposition of Antracol leads to the formation of propylene-thiourea; nothing is known about side effects or about the fate of Antracol and its derivates in algal cultures. In order to avoid any health risks, the application of pesticide to control parasite infections should be restricted to inoculum cultures only.

The occurrence of the parasite reduces the yield of the algal culture but does not represent any health hazard if the infected algal biomass is used as feed or food.

Detailed chemical and biochemical analyses of *Aphelidium*-infested samples of *Scenedesmus obliquus* have been performed in Germany. No mycotoxin that might have been synthesized by this fungus could be detected, and it can be assumed that this parasite does not represent any health hazard if present in algal biomass.

Bacteria

Because commercial algae cultures usually have large exposed basins, the cultures are always contaminated with certain types of bacteria. Heterotrophic bacteria are even found in algal cultures grown autotrophically in inorganic media, because algal cells exude several organic compounds into

Table 10.2 *Effects of different pesticides on* Aphelidium *and* Scenedesmus

Agent	Concentration of active ingredient (mg l^{-1})	Effect on growth	
		Aphelidium	*Scenedesmus*
Afugan	2.0–10	+	+
Afugan	0.2–1.0	0	0
Antracol	10.0–50	+	+
Antracol	2.0	0	**
Aralen	6.0–12	**	0
Calirus	2.0–10	**	0
Calixin	2.0–10	**	0
Daraprim	1.0	0	*
Desinfect	20.0	0	**
Dimanin	0.2–10	+	+
Dithane M 45	10.0–50	+	+
Dithane M 45	1.0	0	*
Flagyl	10.0	**	*
Furalcyn	20.0	0	0
Karathane WD	3.3–13.3	0	**
Karathane LC	0.7–3.3	0	**
Manzate D	2.0–10	0	**
Mertect	3.0–5	**	**
Tecto 60	2.0–10	*	**

Note:
Effects on growth: 0 = no effect; * = weak inhibition; ** = marked inhibition;
+ = cells died.

the medium that are likely to promote bacterial growth, and the degree of contamination is still higher when organic or waste materials are added as nutrients. Because possible public health hazards due to such contaminations will be a point of concern, sanitary safety is one of the important prerequisites for the utilization of algal biomass. Although several studies on the nutritional quality and general toxicity of algae biomass have been published, there are hardly any reports on microbial contaminations that may affect the quality and safety of algae, if they are intended as food or feed.

Most of the microbiological techniques used for the control of food items are specific to the contamination source and not the food type and hence can be applied also for quality testing of microalgae.

Somewhat different is the situation with the standard plate count (SPC). Because the conditions for microbial growth are quite favorable in algae cultures (high levels of dissolved organic matter, high temperature), high SPC levels (upto 10^6 ml^{-1}) can be found with significant seasonal temperature-dependent changes even at the same site.

The most common microbial pathogens found in waters of industrialized countries are *Escherichia coli*, *Salmonella* spp., *Shigella* spp. and *Leptospira* spp., the analyses for which are well established.

Salmonella spp., *Shigella* spp. and *Clostridium perfringens* are the pathogens that in more than 50% of cases account for food-born illness in general. Testing of algal samples of various origin has yet to reveal the occurrence of these organisms and there seems to be peculiar hazard of algal powder contamination by enteric bacteria. The presence of large numbers of *Staphylococcus* spp. indicates questionable sanitary cultivation practices, and special attention is necessary if *Staphylococcus aureus* is found. No doubt the chances for contamination of the liquid culture with this organism are small, as *S. aureus* does not survive in *Spirulina* cultures; there are possibilities for human skin contact and hence contamination of the algae products during the different processing steps.

With the intention of collecting some preliminary data, samples of algae cultures and dried algae powder were screened for bacteria, yeast and moulds (Mahadevaswamy & Venkataraman, 1987). Standard procedures were applied for enumeration by the 'most probable number' (MPN) technique, the number of colonies of each type expressed as 'colony forming unit' (CFU). Among different types of bacteria, coliforms (especially *Escherichia coli*) are considered to be indicative of a sewage or fecal pollution and the possible presence of pathogens. Further isolation of the microorganisms was made by transferring typical colonies from agar plates to nutrient broth, purified on agar and cultured for further biochemical tests. A total of about 100 organisms isolated from algae cultures and dried powder were subjected to biochemical characterization and classification.

Scenedesmus

To study the bacterial infections in *Scenedesmus obliquus*, the algae were raised outdoors under standard growth conditions with inorganic and organic media, and dried by drum-drying. The microbial counts of these different cultures, taken at intervals of three days, are shown in Table 10.3. The initial count (CFU) was 0.3×10^5 and increased to 3.0×10^6 within six days, indicating that the algae culture did favor the bacterial growth probably because of excretion of bacterial promoting compounds. However, this stimulation differs from cases to case; there are reports on pilot-plant cultivation of algae, where no significant increase in bacterial load was observed, indicating that still other parameters besides organic excretion products influence the occurrence of bacteria. Coliforms were found only in a few batches of algae tested and it seems that this type of contamination cannot be avoided completely in algae mass cultivation.

In heterotrophic algae cultures yeast and moulds were found initially at a rate of 150 colonies per milliliter, which increased up to 2000 per milliliter after six days. The wet slurry coming out of the centrifuge is highly

Table 10.3. *Microbial load in cultures of* Scenedesmus obliquus *(tank volume 1400 l)*

Cultural conditions	Age of culture (d)	Dry weight (mg l^{-1})	CFU[a] ml^{-1} ($\times 10^5$)	Yeasts and moulds colonies ml^{-1} ($\times 10^2$)	Coliforms MPN[b] ml^{-1}
Autotrophic ($3 l CO_2 h^{-1}$)	Initial	80	0.3 (0.2–4)	1 (0.3–2)	Nil
	3	200	18 (8–25)	5 (3.8–8)	29 (11–75)
	6	320	30 (25–60)	15 (7–25)	280 (210–460)
Heterotrophic (100 mg of molasses l^{-1})	Initial	80	0.3 (0.2–4)	1 (0.3–2)	Nil
	3	170	12 (5–20)	1.8 (1–6)	120 (75–210)
	6	260	45 (25–80)	5 (4–10)	180 (120–240)
Mixotrophic	Initial	80	0.3 (0.2–4)	1 (0.3–2)	Nil
	3	290	30 (20–80)	10 (3–25)	300 (240–460)
	6	480	400 (200–500)	30 (20–60)	900 (460–1100)

Notes:
Values represent the mean of five independent observations.
[a] CFU, Colony forming unit;   [b] MPN, most probable number.
Figures in parenthesis indicate the range of values.

concentrated, compared with the dilute cultures. As expected, the bacterial load per volume is therefore higher than in the culture. Somewhat surprising was the observation that the bacterial counts are higher in cultures grown with CO_2 and molasses compared with cultures grown on molasses alone. This observation confirms the assumption that the algae themselves enhance bacterial growth, otherwise there should be no significant difference between the two samples with the same amount of molasses in the culture medium.

Table 10.4 shows the list of major species of *Bacillus* and *Micrococcus* that were isolated from outdoor algal cultures; 62 species of *Micrococcus* and 32 species of *Bacillus* could be identified. Total counts of bacteria, yeast and moulds in the drum-dried and sun-dried algae powder are listed in Table 10.5. As expected, the different methods of drying affect the total counts substantially. The lowest loads were in algae powder obtained from

Table 10.4. *List of microorganisms isolated from outdoor algae cultures (percentage occurrence of each species out of 100 randomly isolated colonies)*

	Number of isolates (%)	
Organism	*Scenedesmus*	*Spirulina*
Micrococcus urea	10	—
Micrococcus roseus	10	—
Micrococcus conglomeratus	10	—
Micrococcus flavus	5	—
Micrococcus cryophilus	5	—
Micrococcus freudenreichii	8	—
Micrococcus varians	6	—
Micrococcus caseolyticus	8	—
Bacillus subtilis	5	10
Bacillus coagulans	8	—
Bacillus marcarans	3	5
Bacillus megaterium	2	—
Bacillus lentus	4	15
Bacillus brevis	2	—
Bacillus alvei	1	8
Bacillus circulans	4	10
Bacillus licheniformis	3	20
Bacillus firmus	—	15
Bacillus polymyxa	—	5
Unidentified forms	6	12

Table 10.5. *Microbial load in algal powder dried by different methods*

	Colony forming units g^{-1} ($\times 10^4$)		Yeasts and moulds colonies g^{-1} ($\times 10^2$)	
Method of drying	*Scenedesmus*	*Spirulina*	*Scenedesmus*	*Spirulina*
Drum-dried,	2	3	1	5
autotrophic culture	(0.8–3)	(2–9)	(0.5–1)	(2–10)
Drum dried,	10		25	
heterotrophic culture	(5–15)	—	(10–30)	—
Sun-dried	120	40	150	12
	(100–180	(20–90)	(100–350)	(8–20)
Spray-dried	—	8		5.5
		(2–20)	—	(1–10)
Cross-flow air-dried	—	50	—	20
		(40–100)		(10–409)

Notes:
Values are the means of five independent observations.
Figures in parenthesis indicate the range of values.

Table 10.6. *Microbial counts of drum-dried* Scenedesmus *powder, stored at room temperature*

Storage time (d)	CFU g^{-1} ($\times 10^{-3}$)	Yeast and moulds (colonies g^{-1})
Initial	9	600
15	7	550
30	7	550
45	6.5	500
60	6	450
90	5.5	400

Notes:
Values are means of three independent observations.

drum-drying, whereas sun-dried powder had maximum counts owing to low-temperature treatment. If the cultures are raised with molasses as an additional carbon source, the microbial load in the dried product is slightly higher, even after drum-drying. This observation may be because of an increased amount of spore-forming bacteria that survive the heat treatment when molasses is added as substrate.

Table 10.6 indicates storage studies conducted for a period of three months using drum-dried *Scenedesmus* sealed in aluminum bags. It was found that the microbial load does not increase during storage. This may be explained by the low moisture content (6–8%) of the algae powder, which prevents the growth of microorganisms. As the lipid content of algae normally is not very high, there appears to be less risk of autooxidation leading to rancidity of the product. Whether the algal powder is to be kept in an inert atmosphere such as CO_2 or N_2 depends on the extent to which the algae are produced and marketed.

Spirulina
Similar tests on bacterial contamination were also performed with fresh cultures and dried material of *Spirulina* (Table 10.7). Outdoor cultures of this alga had an initial bacterial contamination of 2.0×10^4 CFU ml^{-1}, which increased to 7.0×10^4 ml^{-1} after 10 days.

Biogas effluent obtained after digestion of poultry droppings in anaerobic digesters contained high bacterial load, yeasts and moulds, anaerobic bacteria and also coliforms. With the growth of *Spirulina* in the culture the microbial load decreased from 60×10^5 to 8×10^5 CFU ml^{-1} during the growth period of 10 days. A similar trend was observed for the anaerobic counts, which also decreased with increase of algae growth. There could be

Table 10.7. *Microbial load in cultures of* Spirulina platensis (*tank volume 2000 l*)

Culture conditions	Age of culture (d)	Dry weight (mg l⁻¹)	CFU ml⁻¹ (× 10⁴)	Total anaerobic counts colonies ml⁻¹ (× 10³)	Yeasts and moulds colonies ml⁻¹ (× 10²)	Coliforms MPN ml⁻¹
Inorganic medium	Initial	150	2 (1–8)	4 (1–7)	1.5 (3–5)	Nil
	5	450	5.5 (4–9)	2 (3–8)	2 (1–8)	Nil
	10	1050	7 (3–10)	0.3 (0.1–0.6)	1.6 (1–5)	Nil
2% Biogas effluent	Initial	150	60 (30–90)	70 (50–90)	400 (150–800)	11.000 (1400–24000)
	5	400	40 (10–60)	8 (2.5–9.0)	90 (45–300)	1800 (460–4600)
	10	950	6 (5–10)	0.5 (0.2–1)	5 (2–9)	Nil

Notes:
Values are the means of five independent observations.
Values in parentheses indicate range of five independent observations.

two possible reasons for these reductions: a) the high alkalinity of the medium, and b) the possible inhibitory role of the leaching from algae biomass. The second possibility was supported in studies involving recycling of the medium where lower bacterial counts were recorded compared with 10-day culture medium and studies on the survival of inoculated pathogens in *Spirulina* cultures. *Escherichia coli* and *Staphylococcus aureus*, when inocculated into the medium, showed a sharp decline and did not survive after six days. In the algal nutrient medium without algae these two pathogens also declined but at a slower rate. Similar trends were also observed for *Salmonella* and *Clostridium perfringens*.

On an average, the counts in *Spirulina* cultures are one order of magnitude less than in *Scenedesmus* cultures. This difference probably may be attributed to the high alkalinity of the *Spirulina* culture medium.

The nature of the various *Bacillus* species found in the algae powder were similar to those identified in the cultures; certain species of *Bacillus* were resistant even at high-temperature treatment. On the other hand, this study has shown the total absence of any pathogenic forms both in the culture and the dried powder. Summarizing the preliminary findings, it has to be stressed that regular monitoring of algal cultures, and particularly of the dried material, is necessary because the microbial load is likely to affect the quality and the safety of the final product.

Reliable comparable data about microbial loads in algae cultures or dried products of other algal species could not be found in the literature and therefore it is not possible to decide whether the contamination described above are also common in other samples. If no special stipulations on the permitted amount of CFU in algae are available, which certainly is true in most of the cases, the standard for baby food may be taken as reference, which is 5×10^5 CFU g^{-1} in several countries. In any case, more-detailed studies are required on maintaining the quality of algae over a long period for extending shelf life and also for studying any deterioration in taste.

Zooplankton

There are occasional reports on contaminations of cultures of green algae with zooplanktons. Sporadic infections with *Lycrymanis* sp., *Colpidium* sp. and *Vorticella* sp. have been described, but these organisms have a negligible effect on algal growth. Contaminations with a group of rotifers called *Branchionus* can be more serious. Heavy growth of these organisms may impair the growth of the algal cells and in extreme cases can even spoil the complete algae culture.

The most effective way of controlling this has been to lower the pH of the culture to about 3.0 by addition of acid and allowing the culture to stand at this pH for 1–2 h. Following this, the pH is readjusted to 7.5 with KOH. This treatment does not affect the algal cultures but very effectively eliminates the rotifers. This procedure is, however, restricted to smaller cultures only because of the amount of chemicals needed to induce the required pH changes.

Contamination with this zooplankton has not been reported for *Spirulina* cultures, probably because of the high pH and salt concentrations in the medium for this alga.

Problems with contamination by zooplankton have been reported from the high-rate algae ponds at Singapore. The most serious problem encountered there was the frequent appearance of the algal predator *Moina* sp. (*Phyllopoda*). Blooms of *Moina* resulted in rapid depletion of the algal population in the ponds. Elevating the pH to above 9 was effective against the predator, but affected the algae and reduced the productivity. Continuous mixing of the culture appears to be effective in controlling predation and improving pond stability.

Insects

Among the insect contaminations, *Chironomus* in *Scendesmus* and *Ephydra* in *Spirulina* cultures have to be mentioned here, as they occur at different algal projects. The eggs of the *Chironomus* insect hatch into wormlike, reddish larvae, which can be found in large numbers during certain seasons on the bottom of the algae ponds (*Scenedesmus*). It is obvious that once the larvae have entered the cultures, they can grow unhindered because of abundant supply of food (algae) and the lack of natural enemies or predators in this artificial biotope. Because of the fact that these larvae a) feed on microalgae, b) use algae and other organic and inorganic material to build protection cases, and c) excrete metabolic wastes that will promote flocculation of algae and bacterial growth, they negatively effect the productivity of the algae cultures. The easiest method to keep this infection within certain limits is to remove the eggs of the insect from the culture before the larvae hatch. This has been done successfully by inserting fine-mesh wire sieves into the current of the agitated algae culture and filtering the mucous egg threads out of the medium.

Contamination of *Spirulina* cultures were noticed during the rainy seasons in tropical regions, when *Ephydra* frequents the algae culture. It was found that the continuously agitated cultures were visited by fewer insects than occasionally stirred cultures. Unlike *Chironomus*, these insects glide on the surface of the culture and lay their eggs into the culture. In the course of detailed investigations, synchronization between algal growth pattern and morphogenesis of *Ephydra* was observed. The 10-day *Spirulina* growth period was found to be adequate for final stages of pupae to hatch as imago. The eggs, laid on the culture surface, reach larval stages in two to three days. During these stages, which include three instar steps, the larvae feed on the algae biomass before they enter the pupal stage, which lasts for two to three days. The reduction of yield due to heavy infections was about 40%. Because the contamination of algae cultures with insect larvae can be rather high during the monsoon in these regions, several control measures were attempted. Biological controls such as fish are not possible as they would not survive in the alkaline *Spirulina* culture medium. Therefore several pesticides were tried for combating the contamination with the insect larvae. Among the chemicals tested, BHC at a concentration of 5 p.p.m. was most effective. Because the application of these chemicals may lead to problems of contamination and accumulation, this method may not be advisable in practice. The use of 20-mesh nylon cloth as a cover over culture ponds proved to be very effective but is not practical for cultivation in large culture areas. At present, no effective means of protecting *Spirulina* cultures from contamination with *Ephydra* larvae can be recommended.

One has to ask whether the presence of fragments from indigenous insects or arachnides, summarized under the expression 'filth', in the dried algae material really represents a health risk and/or affects the aesthetic properties.

Because almost all commercially produced algae are cultivated in open ponds they are susceptible to contaminations by insects. Hence, it is difficult to establish a common standard because of varying culture and processing conditions, in which insect residues can be fragmented into particles so small that they cannot be separated out by standard methods.

Another kind of contamination that can be found in algal samples is residues from plant material, which normally is ignored as its contribution to the algal biomass is insignificant. However, because the fibrous plant material is often difficult to distinguish from insect particles, it is often added to the category 'unidentified insect fragments', thus erroneously overestimating the amount of this contaminant. To overcome these difficulties, an interim first-action flotation method was developed by Nakashina (1989) to detect light filth in algae powders and tablets, which was adopted final action by the Association of the Official Analytical Chemists (1990). Although this procedure was designed initially for the detection of impurities in *Spirulina*, it can also be applied for other algae. The method consists of dispersing 50 g of algae material in water, boiling with dilute HCl and wet sieving. The residue is brought to the boil in water with the addition of mineral oil. After stirring, the product is removed by washings in a percolator, the mineral oil is filtered, and the filth is counted.

The details for the procedure are as follows. Mix 50 g of algae with portions of 500 ml tap water, reserving an amount to rinse the spoon, magnetically stir on a hot plate at low setting until all materials are completely dispersed. If necessary, add a few drops of antifoam solution and then add 15 ml of HCl in increments with magnetic stirring to disperse foam: if no foam occurs, add remaining HCl and boil for 30 min with stirring. Clean the beaker sides with a rubber policeman, and gently pour the contents through a sieve no. 230 and rinse with hot water (> 50 °C) until the washings are clear.

Transfer the residue quantitatively with water back to a 1 l beaker and fill the beaker up to 500 ml. Bring mixture to the point of boiling under stirring but do not allow to boil. Immediately add 150 ml mineral oil, stir for 10 min and let stand for 5 min. Add contents to percolator containing about 250 ml water by rinsing beaker and stirring bar with hot (> 50 °C) water. Fill percolator to 1700 ml with water, stir contents with glass rod 2–3 s. Let stand for 3 min and drain percolator to 250 ml level. Repeat the filling and draining until the lower aqueous phase is clear or free of suspended material. Drain the oil layer into retained beaker, then rinse glass rod into beaker with isopropanol, wash percolator sides successively with hot (> 50 °C) water, isopropanol and 1% Na laurylsulfate. Boil beaker contents for 2 min and immediately filter beaker contents and wash beaker successively with hot (> 50 °C) water, isopropanol and 1% Na laurylsulfate and examine the filter paper microscopically at 30 × magnification.

Current FDA guidelines permit up to 300 unidentified insect fragments per 100 g of algae.

Other contaminations

Because open algal ponds represent a large area of almost stagnant water in a sometimes arid or semiarid terrain, they are attractive to different kinds of vertebrates, mainly birds and rodents. Though the animals themselves do not pose a real problem, the presence of rodent hairs or feather fragments in the dried algae are indicators of potential contamination because these parts may carry pathogenic contaminants such as bacteria. According to the current FDA guidelines, one rodent hair per 100 g of *Spirulina* is permitted, whereas hair from humans and indoor-dwelling rodents (mice and rats) should be totally absent.

Therefore, care has to be taken to protect the ponds from these animals; this is easier for rodents than for birds, which are attracted by the insects that always appear near to the ponds.

10.7.   Harvesting

General introduction

To a considerable extent, the economy of microalgae production depends on the technology employed for harvesting and concentrating the algal suspension in order to render a slurry suitable for further processing. Of the many harvesting techniques that were developed and tested during the past three decades, only a few were found to be effective at a reasonable operational cost.

Separation of algae from the medium faces several difficulties: aside from their minute concentration in the culture medium, most algae have a size smaller than 30 μm (with the exception of *Spirulina*) and density only slightly greater than that of water. In practice, this means concentrating the fairly dilute culture (200–600 mg l^{-1} or 0.02–0.06% dry solids) until a slurry, paste or cake containing 5–20% or more dry solids is obtained. The choice of suitable harvesting process depends on the algal species and the purpose for which the algae are required. For instance, chemically induced flocculation may result in a product contaminated with residues of the flocculant, or centrifugation may damage sensitive algal cells leading to the release of cellular materials that may affect the quality of the algal product.

Before coming to a decision as to which harvesting process will be the most suitable one for a particular case, the following general questions should be considered.

1) What type of algae has to be harvested (unicellular-coccoid, unicellular with spines, filamentous)?
2) Does the harvesting device operate continuously or discontinuously?
3) Does the device preconcentrate only, does it concentrate in a single step or does it need preconcentration beforehand?
4) What is the percentage of dry matter of the concentrate?
5) What are the investment costs?
6) What is the energy demand per cubic meter of algal suspension?

In connection with the last point it has to be remembered that some microalgal processes, especially in the field of waste-water treatment, merely require separation of the biomass and do not necessarily involve production of a dry product.

Gravity filtration

The simplest and cheapest method for separating suspended solids of sufficient size from liquid is by gravity filtration such as screening or straining. It is widely used in chemical engineering and has also been applied for the harvesting of algae. This method becomes difficult if the particles are compressible or display a certain plasticity like many unicellular algae such as *Chlorella* sp. or *Scenedemus* sp. Therefore, gravity filtration is restricted to filamentous or colony-forming algae such as *Spirulina* sp., *Oscillatoria* sp. or *Micractinium* sp. because of limitations in the pore sizes of the fabrics available. These latter algal species are able to bridge the openings of screens with pore sizes between 50 and 100 μm so that the medium can drain off to be discharged or returned to the cultivation pond. It is possible to filter *Spirulina* even with ordinary cloth material.

Different modifications of simple cloth filters, all working on the principle of gravity filtration, that are suitable for smaller *Spirulina* cultivation units, are illustrated in Fig. 10.55 *a–d*. Fig. 10.55 *a* shows a two-deck filter, comprising two hemisperical filters, held in a steel frame and draining into a bottom collector. A 25-mesh nylon cloth is used in the upper deck to remove undesired extraneous fibrous material and a 60-mesh cloth for the lower deck to harvest the algae. This deck is equipped with a rubber-padded scraper meeting the contour of the deck; the agitator stem is connected to the rotable upper deck. In order to increase the throughput rate of this separation technique and to achieve continuous harvesting of the concentrated algae, various mechanically agitated screens have been designed, such as vibrating screens, rotating screens or microstrainers. Corrosion of the screen material has been overcome by using plastic materials; clogging by small algae or slimes, however, still remains a problem. Vibrating screens, operating with 300- to 500-mesh fabrics, are used in Israel for the harvesting of *Spirulina* with an efficiency of 95%. A considerable drawback of this system is the rubbing effect on the algal filaments, causing rupture of the cells, leakage of cell content and subsequently an increase of the organic load of the pond if the filtrate is reused.

For the fragile alga *Dunaliella*, several gentle harvesting methods have been tested, for instance passing the dilute algal culture through diatomaceous earth and extraction of the desired lipophile ingredient (β-carotene) with organic solvent from the earth. The drawback of this system is the low efficiency of filtration and the need to wash the earth from the remaining algal debris.

Microstrainer

Microstrainers, designed initially for the elimination of particulate matter from effluents of sewage-treatment plants, have also been used for the removal of algae. The device is reliable for the harvesting of filamentous algae, but unsatisfactory for the removal of smaller algae that cannot be retained on the 20 μm fabric used.

A potential disadvantage associated with all these harvesting techniques is the fact that they preferentially remove the filamentous larger algae from the pond. This may result in an enrichment of the medium with small contaminating algal species, which after several harvesting steps may even overgrow the initial algal species. Green algae such as *Chlorella* and *Scenedesmus* pass through these screens with no removal; only openings smaller than 20 μm will retain *Scenedesmus* sp. and openings of 5 μm are required to retain most *Chlorella* species. On the other hand, these fine screens have very low throughput rates with less than 10 l m^{-2} min^{-1} so that continuous backwashing is required to prevent clogging.

Lamella separator

An improved harvesting device based on sedimentation at natural gravity is the lamella separator, which offers an increased settling area by the combination of a number of sloped plates (Fig. 10.56). The slopes ensure a downward movement of the sediment into a sump from where it can be removed. The algal suspension is pumped continuously whereas the slurry is removed discontinuously. The average concentration reached for *Coelastrum*, for instance, was 1.6%, good enough for preconcentration. This type of harvester is also suitable for filamentous algae; for the harvest of unicellular algae flocculants have to be added, which increase the operation expenses.

New fabrics now available down to about 1 μm nominal rating allow capture of small unicellular algae albeit at a low filtration rate. Thin filter cake, water carryover at higher drum rotation speeds and the use of water jets to dislodge the cake cause severe dilution of the removed solids resulting in concentration factors of only ten times.

Belt filter

For applications where complete recovery of unicellular algae is an objective, a greater concentrating factor than that obtained with ordinary microstrainers is required for economic feasibility, although the gravity filtration principle employed by microstrainers is advantageous particularly where utilization of the algal biomass as feed or food is contemplated.

The use of flocculating chemicals is mainly avoided because that would add to operating costs and also may contaminate the product. To increase the concentration efficiency and also to retain small algal cells, precoated belt filters for harvesting algae were tested at several places. For instance, a paper-fiber precoat was used as the filter medium, deposited on a coarse mesh polyester fabric belt. After deposition of the algal cells on the precoat, the algae–paper mat was dewatered and peeled off continuously from the belt. The major limitation of this process is the fact that the final product, which is a mixture of paper filters and algae, is only suitable as feed for ruminants.

As a result of the experiences with precoated filters, and the introduction of fine-weave fabrics, a new type of continuous fine-weave belt filter has been developed for efficient preconcentrating of all types of algae. This device, which works as a reverse microsieve, is based on the gravity filtration principle but recovers the algal solid cakes by suction rather than by water backwash, thereby providing higher concentration factors (Dodd, 1979). The device consists of a large filtration drum, a much smaller separation drum for suction recovery of solids, a polyester fabric belt with

Fig. 10.55. *a*, *b*, Simple gravity cloth filters for the separation of *Spirulina* (CFTRI, Mysore, India); *c*, rural type of a two-deck filter for the separation of *Spirulina* (MCRC, Madras, India); *d*, sloped two-step filter unit for the separation of *Spirulina* (Taiwan).

c

d

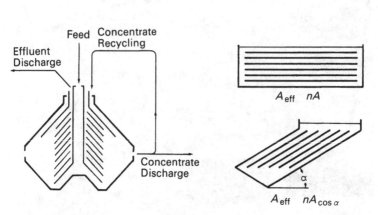

Fig. 10.56. *a*, Schematic cross section of a lamella seperator. *b*, Derivation of the working principle of lamella separators, *nA*, Number of sedimentation surfaces; A_{eff}, effective sedimentation surface.

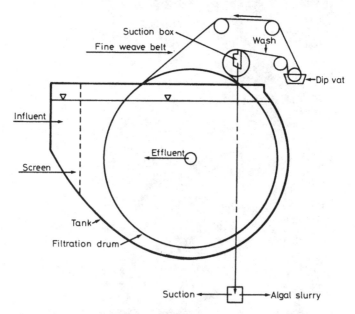

Fig. 10.57. Schematic diagram of a fine-weave belt filter (after Dodd, 1986).

12 or 5 μm nominal pore size, a storage tank, belt rollers and washing spray bars. A schematic diagram of this harvester is shown in Fig. 10.57.

The suspension to be separated flows from the outside into the large drum covered with the rotating filter fabric. Because the liquid level outside the sieve drum is greater than in the drum, a differential pressure results, forcing the medium through the fabric. By rotating the drum, the filter cloth with the algae adhering to it ends up outside the suspension where the concentrated algal biomass is sucked from the cloth through small slits. After that, in a further step, the fabric is cleaned with spray water. The performance of the system was quite satisfactory when certain precautions were observed. For instance, a loss of throughput may occur if the belt is not cleaned after each run, owing to pore clogging if deposits are allowed to dry on the belt. It was found that hypochlorite washing using the dip vat at the tensioning roller was the most effective method of cleaning. The efficiency of the harvester can be expressed based on flow throughput (m^3 m^{-2} drum surface h^{-1}), recovery rate expressed as percentage of total suspended solids reduction, and slurry content expressed as percentage of total solids.

Experiments performed with different algal species showed that with a filter material having 12 μm pore size, a throughput up to 17 m^3 h^{-1} can be obtained at a belt speed of 22 m min^{-1}. A linear relationship between throughput and belt travel speed was found when the results were plotted on a log–log scale as shown in Fig. 10.58. The slurry concentration, however, is inversely proportional to the belt speed and ranged between 1.5 and 3.0% solids. Power requirements averaged between 0.3 and 0.5 kWh m^{-3} of algal suspension harvested, giving a 40-fold increase in slurry concentration from about 0.05% to 3%.

The most difficult problem encountered with this type of harvester is linked to the size of the algae harvested. Filamentous or coenobial cells such as *Spirulina* sp., *Micractinium* sp. or *Scenedesmus* sp. are ideal for separation with such harvesters (Fig. 10.59), whereas unicellular forms such as *Chlorella* or *Oocystis* are more difficult to retain on the belt. They easily block the pores of the fabric and any pressure will cause the cells to go through the pores. These algae could be harvested with a 5 μm pore fabric; however, with this fine material, the throughput capacity of the machine dropped to about 2–3 m^3 h^{-1} (Fig. 10.60).

In addition to the various filtration techniques that are based on gravity, several other devices have been developed by which the filtration efficiency is increased either by applying vacuum or pressure.

Vacuum drum-filter

Vacuum drum-filters with precoated surfaces are frequently used to recover yeast from fermentation broths (Fig. 10.61). Because the size of several microalgae such as *Chlorella* sp. or *Scenedesmus* sp. are in the same order of magnitude as that of yeast cells, this harvesting technique is also applicable for these algae. The operational principle is as follows. As an initial step, a trough, surrounding the lower part of the filter drum, is filled with a suspension of starch grains (potato or corn). By creating a vacuum inside

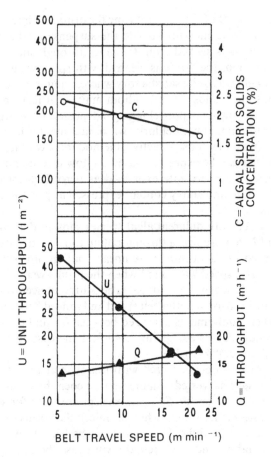

Fig. 10.58. Performance of belt filter at various belt travel speeds. Belt material was polyester fabric of 12 μm porosity, belt width was 0.95 m.

the drum and by turning the drum, a starch precoat of about 2–3 cm thickness is built up on the outer filter surface. Thereafter the trough is filled with the algal suspension so that an algal filter cake will be added upon the precoat. By slowly moving a blade towards the rotating drum, the algal filter cake is removed together with a small outermost part of the precoat, thereby exposing fresh precoat surface. With this system, throughput rates of about 0.5 m^3 m^{-2} filter h^{-1} can be achieved. To obtain optimal results it is necessary to use preconcentrated algal suspensions of about 2% dry matter. The disadvantage of this process is the fact that the harvested algae contain a certain amount of precoat material (15–30%), which may limit the further utilization of the product.

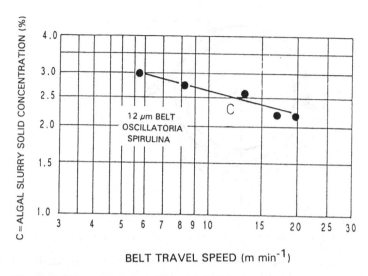

Fig. 10.59. Filter performance with large-celled algal species.

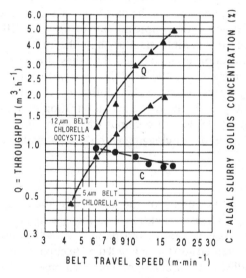

Fig. 10.60. Filter performance with small algal species requiring narrow filter-belt material with a pore size of 5 μm.

Chamber filter press

The most common filtration device using additional pressure is the chamber filter press. For separating algae, filter presses are required with chamber gaps of less than 10–15 mm and filtration pressures of up to 16×10^5 Pa as well as filtration cycles of 1–2 h. For easier removal of the

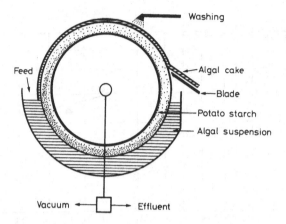

Fig. 10.61. Schematic design of precoated vacuum drum-filter.

filter cake, diaphragm presses have been developed, with plastic or rubber diaphragms mounted behind the filter cloth. Whereas in the case of conventional presses the harvested material is removed by continuously pumping suspension into the chamber, in the new type the cake is pressed off by inflating the diaphragm with compressed air of up to 6×10^5 Pa. These new machines are considerably smaller than the previous ones because of the shorter filtration cycles needed, but they are also more expensive. It has been reported that 100% of *Scenedesmus* sp. could be harvested with a diaphragm press when the filter cloths were covered with a potato starch precoat (500 g m^{-2}) before the actual filtration process (0.6 m^3 m^{-2} h^{-1}). This comparable slow throughput was chosen in order to prevent penetration of the precoat or clogging with the algal material.

Flocculation

During the logarithmic growth phase of the algal cells, their negative surface charge is high and difficult to neutralize so that the algae remain dispersed. During longer residence times and reduced growth rates, the negative charges will be decreased and the cells will tend to clump and settle, resulting in a process called autoflocculation. This phenomenon can be accelerated by organic polymers excreted by the algae into the medium.

If such a method of algal separation is envisaged, one may give up much in the way of productivity. A decision has to be made between either fast-growing cultures, which require costly harvesting processes, or less-productive ones, which are easier to harvest; i.e. a compromise has to be found where the loss of productivity is offset by decreased costs of harvest.

As mentioned before, suspensions of microalgae are so stable under optimal growth conditions that spontaneous sedimentations are negligible. If, however, the conditions are not favorable, algal growth rate will decrease thereby causing apparent spontaneous floc formation and sedimentation of the algae. The phenomenon whereby algae spontaneously flocculate and settle is well known from both the ecological and sanitary engineering literature. The factors and processes by which algae are induced to flocculate have not yet been satisfactorily elucidated; however, nutrient limitation or high photosynthetic activity at limited CO_2 supply seem to be important parameters. In some cases, this autoflocculation is associated with elevated pH due to CO_2, nitrate and phosphate assimilation and coprecipitation of magnesium, calcium and carbonate salts. In contrast to this process, commonly called autoflocculation, the term bioflocculation is used to denote algal flocculation that is not caused by coprecipitation with mineral salts.

The formation of algal aggregates can also be induced by interactions between algae and bacteria or excreted organic macromolecules. Studies performed on the mechanisms of autoflocculation indicate that this method can be initiated by certain steps in algal pond management. Stopping pond agitation and/or CO_2 supply cause an increase in pH leading to the formation of algal flocs, which will settle to the bottom of the pond. Experiments performed with *Scenedesmus* sp. on the effect of different pH values showed that no flocculation could be obtained for values between 5.0 and 7.5, whereas at pH values above 8.5 almost 95% of the algal biomass could be removed. Another contributory factor for the induction of bioflocculation is the concentration of magnesium, calcium and phosphate in the medium. Free calcium phosphate precipitate has a positive surface electric charge and therefore may be absorbed by the algae to neutralize their negative surface charge and thus promote algal flocculation.

From Peru it was reported that in the hard water available there, the addition of ammonia in amounts required to replenish the nitrogen consumed by the algal biomass was sufficient to get a nearly complete flocculation of *Scenedesmus* sp. within a few minutes. The efficiency of this flocculation was improved when the amount of phosphate required for fertilizing the culture was added as phosphoric acid immediately before the addition of ammonia.

The effective ion of this process is not ammonium but the calcium ion, present in concentrations between 100 and 160 mg l^{-1}. Some experimental ponds have been constructed that make use of this very simple separation technique and allow the harvest of algae settled due to autoflocculation (Fig. 10.20). These ponds, however, often have the disadvantage of requiring special construction. Moreover, they are not available for algal cultivation during harvesting and this results in considerable additional space requirements.

For smaller algal cultivation units, tapering sedimentation tanks or vertical sedimentation tubes with a conical bottom part have been described giving solid fractions of 1.5%. To avoid penetration of the supernatant into the sediment during removal of the settled algae, a withdrawal pipe is recommended, which reaches through the tank or tube into the bottom part and through which the sediment can be sucked up by means of a pump.

Because the process of autoflocculation, as described before, occurs only under certain and not always predictable environmental conditions, this method of algal preconcentration is not very reliable. To overcome this problem, several techniques have been developed to initiate the algal sedimentation by the application of various flocculants. This method is widely used in water and waste-water treatment. The flocculants commonly used include several polycations such as aluminum and iron salts. The typical reaction of these compounds are:

$$Al_2(SO_4)_3 + 3Ca(HCO_3)_2 \longrightarrow 3CaSO_4 + 2Al(OH)_3 + 6CO_2$$
$$Fe_2(SO_4)_3 + 3Ca(HCO_3)_2 \longrightarrow 3CaSO_4 + 2Fe(OH)_3 + 6CO_2$$
$$2AlCl_3 + 3Ca(HCO_3)_2 \longrightarrow 3CaCl_2 + 2Al(OH)_3 + 6CO_2$$
$$2FeCl_3 + 3Ca(HCO_3)_2 \longrightarrow 3CaCl_2 + 2Fe(OH)_3 + 6CO_2.$$

The metallic hydroxides formed are insoluble and tend to clump together, entrapping the algal cells while they settle or float. Besides these metallic flocculants, the residues of which may lead to toxicological problems in the harvested algal biomass, other agents have been recommended for initiating sedimentation that are generally recognized to be toxically safe, such as potato starch derivates, chitosan or synthetically produced flocculants.

With the intention of finding the most economical coagulant for the flocculation of algae, several chemicals were tested by McGarry & Tongkasame (1971) including aluminum compounds, lime, ferrous sulfate, ferric chloride and 50 commercially available polyelectrolytes. By considering primary coagulant dose, pH, mixing conditions, and time of addition, it was found that among the inorganic primary coagulants, alum appeared to be the most feasible. The most economic application was the addition of 150 mg of alum per liter at a pH of 6.5. Even better settling and supernatant clarity was achieved at pH 5.5. However, this is more expensive because of the additional acid required for lowering the pH. Other reports found an optimum pH for Fe^{3+} flocculation at about 3.5 and doses in the order of 50 mg $FeCl_3$ l^{-1}. Optimum pH for alum flocculation was 5.3–5.6 and optimum dose in the range of 80–250 mg l^{-1}. The polyelectrolytes were screened either as primary coagulants or in combination with alum. Several cationic compounds were found to be effective, whereas anionics and non-ionics induced flocculation only at excessively high doses. The best results were achieved with polycations, but the amounts were equivalent to that of alum. In mixtures of alum and polyelectrolytes up to 3 mg l^{-1}, the usage of the latter decreased the costs and reduced the amount of alum required by 60 mg l^{-1}. Because residues of alum (average value 4% w/v) in the final product may cause toxic side effects, the use of polyelectrolytes could provide benefits in terms of lowering the alum content in the harvested algae.

Different experiments on the artificial precipitation of algae were reported from China (Yu *et al.*, 1959). In mass cultures of *Scenedesmus obliquus*, HCl, KOH, NaOH, $Al_2(SO_4)_3$, CaO and saturated lime water $(Ca(OH)_2)$ were tried as flocculants. Lime flocculation is a common technique in water and waste treatment and involves raising of the pH to the point at which $Mg(OH)_2$ is formed to act as the ultimate flocculant.

Among the flocculants tested, HCl, KOH and NaOH were not very effective; the algae precipitated very slowly and became yellowish. Treatment with 0.03–0.05% of alum or with 6% of lime water resulted in the complete precipitation of the algae within 2 h. As with several other microalgae, alum at concentrations of about $150 \, mg \, l^{-1}$ was found to be the most efficient chemical flocculant for *Dunaliella*.

In general, the flocculated algae cannot be used directly for food or feed purposes unless the flocculant is absolutely harmless or is removed prior to further utilization. Adhering flocculant on the surface of the *Dunaliella* cells does not cause serious problems, if the alga is cultivated for the production of β-carotene, which is removed by extraction with organic solvents, thus leaving behind the flocculant-contaminated algal debris.

In order to overcome these problems in general, a new principle of harvesting *Dunaliella*, developed in Australia, has been mentioned by Borowitzka & Borowitzka (1989). The method is based on the finding that this alga, if cultivated in media with salt concentration above 4 M, acquires a hydrophobic surface coat that aids the adsorbance of the algae to a hydrophobic surface, from where it can be extracted. The major limitation of this possibility of harvesting is the observation that this process does not work with actively growing cells but only during the phase of transition of a motile cell to a non-motile cyst, which seems to be linked with the formation of the hydrophobic cell coat.

A novel flocculation technique using colloidal gas aphrons (very small gas bubbles in the order of 25 μm in diameter encapsulated in a surfactant film) has been investigated for harvesting unicellular algae (Honeycutt, Wallis & Sebba, 1983). The method gave promising results in laboratory trials but has not been tested yet in large outdoor ponds.

Alternatives to the flocculants mentioned above could be potato starch derivates or the cationic polymer chitosan (deacetylated polymer of β-*N*-acetyl-D-glucosamine), prepared by acid hydrolysis from the exoskeleton of marine crustaceans. This biological carbohydrate is been used frequently for the clarification of drinking water and in the food industry and does not impart any toxicity to the products. Concentrations of 50 mg chitosan per liter are sufficient to obtain almost complete (96%) settling of *Scenedesmus obliquus* between pH 7.5 and 8.5. The flocculant is not effective in concentrating *Spirulina*, but this is irrelevant as this alga can be harvested by filtration.

The utilization of chitosan offers some advantages in comparison with that of other conventional flocculants. There has been no evidence of toxic side effects of chitosan; it can be used in very low concentrations and it can be produced easily. The chitinous shellfish wastes are separated from protein residues by treatment with 2% NaOH at 65 °C, the calcium components are removed by extraction with hydrochloric acid and the remaining chitin is converted to chitosan by deacetylation with 50% NaOH at 130–150 °C.

A linear relationship was found between the ionic strength of the medium and alum optimal dosages for algae flocculation. The ionic strength (I) of a given medium is calculated by the equation

$$I = 0.5 \times \sum C_i Z_i^2$$

where C_i is the molar concentration of the ith ion and Z_i its valency. The results indicate that high salt concentrations inhibit the flocculation process evidently by both reducing the chemical activity of the flocculant and by masking its functional active sites. Consequently, the inorganic flocculation demands increase as the medium strength is increased.

A similar effect can also be observed with organic flocculants such as chitosan. Chitosan is effective as a flocculant only when the ionic strength of the culture medium is lower than 0.1. As the salt concentration increases, the intrinsic viscosity value of chitosan decreases rapidly from 2000 ml g^{-1} to a constant value of 100 ml g^{-1}.

It has been suggested that an oxidation pretreatment would reduce the algal motility, change its surface characteristics and improve its flocculability. Pretreatment by ozone was effective and improved the flocculation process by both increasing the algal removal efficiency and reducing the optimal flocculant dosage. It is suggested that this effect is caused by modifications of the algal surface characteristics.

Cationic polyethyleneamine has been reported to be an effective flocculant for cultures of *Chlorella* sp. The flocculation efficiency increased with the increase in molecular weight of the polymer from 800 to about 2000.

Finally, mention should be made of a point that has been observed in connection with the use of flocculants. It has been reported, especially with alum-induced flocculation, that because of the high water-retention capacity of the colloid, the algal slurry cannot be easily dewatered by decantation or filtration. On the other hand, direct dehydration of such dilute slurries is too costly. Preliminary experiments indicate the possibility of breaking the colloid by heating to 90–100 °C; acidification to pH 3.5 prior to heating improves the rate of filtration resulting in cakes with 14–19% dry matter.

Dissolved air flotation

The principle of flotation is to let tiny air bubbles transport solid particles (algae) in suspension to the surface of a normally open tank where they may be removed by skimming or other means. Generally, algal particles can be made to float upwards much more rapidly than they settle downwards so that the removal of algal flocs floating on the surface is more efficient than the removal of precipitated material from the bottom of the settling tank. In addition, flotation processes achieve higher solid fractions (up to 7%) in the concentrate.

These flocs can be effectively separated by attaching them to gas bubbles, which will cause the combination to float to the surface. The bubbles should be as small as possible and their introduction should not increase turbulence in the fluid; however, for effective flotation it is necessary that each particle requires a certain quantity of air around it to reduce its specific mass to a point where it will rise to the surface. All that is best insured by creating the bubbles *in situ*, either by effervescence from supersaturated solutions or by electrolysis, where small bubbles of hydrogen and oxygen are generated which carry the flocculated algae to the surface. For satisfactory clarification the nominal residence time could be less than 10 min. In dissolved air flotation, recycle pressurized to 3 atm and thus supplying an air/solids weight ratio of about 0.01 is sufficient. Algal float containing 4–6% solids can be obtained by skimming off only that part of the float not submerged in the medium. Schematic diagrams of harvesters based on the principles of dissolved air flotation are given in Figs 10.62 and 10.63.

For higher concentrations, the skimmed float, which contains 40–60% air by volume, has to undergo further drainage (secondary flotation). After standing for 2–4 h, the solid content will increase to 6–8%. For this process, vessels with height-to-diameter ratios of near unity are recommended.

The hydraulic flow pattern of the dissolved air flotator is of primary importance to the efficient functioning of the systems. The following simplified mathematical model can form the basis of the design of an air flotation system:

$$A/S = RC_S(f(P+1)-1)/QS_i,$$

where A/S is the air/solid ratio (kg of air per kg of solids)

Q/R = recycle ratio, i.e. recycle flow rate $(m^3\ h^{-1})$/algal suspension flow rate $(m^3\ h^{-1})$

C_S = saturation concentration of air in water $(l\ m^{-3})$

P = operating pressure (atm)

f = proportionality factor, which is a function of the saturation tank

S_i = influent suspended solids concentration $(mg\ l^{-1})$.

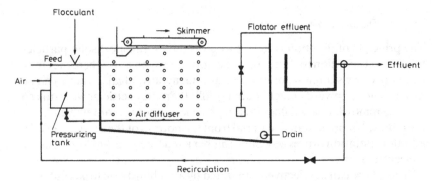

Fig. 10.62. Schematic diagram of harvester based on principles of dissolved air flotation.

A schematic design of a pilot plant scale model is given in Fig. 10.63. Based on preliminary results, the following parameters were calculated: ratio $= 0.02$ to 0.04; $f = 0.3 - 0.6$; $P = 3$ t.

To improve the mixing of the algae with that of the flocculants, a flocculator had been incorporated. It consists of a horizontal cage paddle flocculator running at a very slow speed of about 25 rpm. The addition of the flocculant is dosed by means of a diaphragm metering pump. The flocculator has a reaction time of less than 3 min, which is sufficient to attain stable large flocs before overflowing into the tank.

Centrifugation

The most direct method for thorough removal of almost all types of microalgae is centrifugation. Centrifugation is equally applicable for filamentous and non-filamentous algae. Some algae, such as *Spirulina* sp., which are rich in air vesicles, tend to rise in a centrifuge rather than settle and hence can be harvested more economically with other, simpler systems. Although centrifugation is reliable and simple and yields a product free of added flocculants or other chemicals, high costs for investment and maintenance as well as a high energy demand seem to restrict the employment of centrifuges to such cases, where high selling prices of the algal biomass will offset these expenses.

Among the various types of centrifuges, four major groups can be distinguished, i.e. chamber centrifuges, self-cleaning plate centrifuges, nozzle centrifuges and decanters.

Chamber centrifuge

Chamber centrifuges are suitable where smaller quantities with low solid fractions are to be purified, because this type requires long discharging intervals of about 2 h with comparable slow throughput rates (maximum $5 \text{ m}^3 \text{ h}^{-1}$). The solid concentration achieved is about 20%.

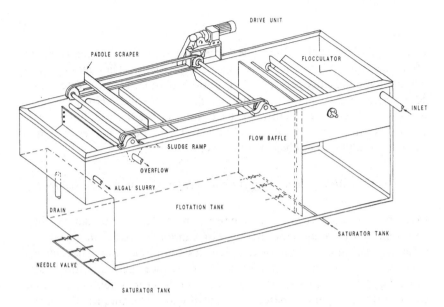

Fig. 10.63. Design of dissolved air flotation harvester.

Plate separators

Self-desludging plate separators are used in many industrial processes for the separation of liquid phases with small differences in density. These machines are employed successfully for harvesting *Scenedesmus* sp. and other types of microalgae from suspensions containing 0.05–0.1% of dry algal matter, giving a concentrate with about 12% algal matter. Plate separators can be operated continuously for long periods of time. The discontinuous discharge of the concentrated solids can be controlled manually or automatically.

Nozzle centrifuge

Nozzle centrifuges, which are common in the yeast industry, resemble the plate separator in many respects. The sludge discharge, however, is continuous through peripherally installed nozzles, resulting in a more homogeneous product. If coarser particles are present in the suspension to be harvested, sieves have to be installed in front of the inlet to remove these particles, which would otherwise clog the nozzles. The concentration of the suspended solids is only by a factor of 10–20 (2% dry solids), which is acceptable if the separator is used for preconcentration only. For obtaining higher concentration factors, part of the primary concentrate has to be recycled without passing the stack of plates again. The advantages of this device as compared with the plate separator are lower investment for the same capacity, and higher dry-matter content of the final recycled concentrate. At a recycling factor of 12 a concentration of 15–20% dry matter could be obtained for *Scenedesmus* sp.

Decanter

Decanters, which achieve the highest solids fraction in the concentrate, are often used for sludge concentration in sewage plants. They require little maintenance and can operate for long periods of time. They operate with a continuous feed of suspension and discharge of solids. However, it is necessary to feed preconcentrated suspensions (1.5–2% dry solids) if high concentration factors should be obtained. This type of harvester is not suitable for all types of algae, for instance *Chlorella*.

Economics of harvesting systems: centrifugation versus dissolved air flotation

In recent publications dealing with the economics of algal production, it has been repeatedly stated that centrifugation cannot be used economically in most cases because the costs are too high, and the application of flocculants is recommended as an alternative means to reduce the harvesting costs. Indeed, at first glance, the use of a flocculant/flotator system seems to be considerably cheaper compared with centrifugation techniques. However, detailed calculations based on results of several field trials demonstrate that the differences in harvesting costs between both methods are very small.

This should be illustrated by a few data, collected in comparative studies on harvesting techniques for *Dunaliella*, i.e. centrifugation versus flocculation. It has to be stressed that the extreme salinity (20%) of the culture medium did not affect either procedure.

Costs for equipment and energy contribute substantially to the production costs of algal biomass. Although the exact percentages of each factor has to be calculated based on the respective local conditions, the following formula is a very useful tool to estimate approximate annual equipment costs including depreciation, interest, maintenance, and energy (Mohn & Contreras, 1991).

$$Y = \frac{(0.5 \times I + M) \times C}{100} + \frac{C}{A} + R$$

Y = annual costs of the equipment
I = interest in percent per year
M = maintenance in percent of capital costs
A = amortization in years
R = running costs: power or chemicals per year

The various basic assumptions and estimated costs for this model calculation on harvesting costs by using the above formula are summarized in Tables 10.8–10.10.

The centrifuge used was a self-cleaning plate separator running as a clarifier with closed rising channels. Other modifications such as using open channels or running the centrifuge as a separator harvested less effectively.

Table 10.8. *General assumptions for the calculation of harvesting costs*

1. Biomass production per year:	25 t
2. Total pond surface:	190000 m²
3. Throughput of the harvester:	10.5 m³ h⁻¹
4. Operation period per year:	330 d
5. Working hours:	18 h d⁻¹
6. Growth rate:	4 g d⁻⁴ m⁻²
7. Harvesting density:	4 g l⁻¹
8. Energy costs:	0.17 US$ kWh⁻¹
9. Capital interest for investment:	16.5% per year
10. Maintenance:	2% of investment per year

Table 10.9. *Harvesting costs by means of centrifugation*

Equipment	Investment (US$)	Amortization (years)	Costs per year (US$)
1. Centrifuge (20m³ h⁻¹)	160000	12	29733
2. Energy for centrifugation	—	—	8530
3. Pump (20 m³ h⁻¹)	2000	12	372
4. Energy for pump	—	—	901
5. Basket filter	3000	12	565
6. Storage tank (1.5 m³)	1000	12	188
7. Mixer for harvested biomass	4000	12	733
8. Energy for mixer			505
9. Storage tank for harvested biomass	4500	12	825
10. Energy for tank			300
Total	174500	—	42652

According to this estimation the average costs to harvest 1 kg of algal biomass by means of a centrifuge are US$ 1.71.

In order to optimize the flocculation/flotation method, several combinations and concentrations of different flocculants were tested for their efficiency. The most promising results of these trials are listed in Table 10.11. A combination of $Al_2(SO_3)_4$ and the anionic polymer (Fig. 10.64) 'Praestol' flocculated the biomass best and the reproducibility was good. The aluminum sulfate could be replaced by $FeCl_3$ as it is less toxic. On the other hand, iron compounds color the medium yellowish, thus rendering it useless for recycling.

Based on recent quotations, average costs for flocculants to obtain 1 kg of algal biomass are US$ 0.61. This price does not include energy for the flotation unit, for instance for air decompression, recycling pump and flotate removal.

Table 10.10. *Harvesting costs by means of flocculation*

Equipment	Investment (US$)	Amortization (years)	Costs per year (US$)
1. Flocculants (100 p.p.m. $Al_2(SO_3)_4$ and 10 p.p.m. Praestol)	—	—	15060
2. Storage tank for alum (2.5 t)	2000	12	376
3. Storage tank for Praestol (0.5 t)	1500	12	283
4. Tank for daily aluminum solution (1%) (2 m³)	7000	12	1318
5. Energy for stirrer			56
6. Tank for daily Praestol solution (0.05%) (0.5 m³)	4000		753
7. Energy for stirrer			56
8. HCl (pH adjustment)			200
9. Tank for HCl	3000	12	565
10. Flotation unit	70000	12	13000
11. Energy for flotator (0.14 kW m⁻³)			1484
12. Pump (20 m³ h⁻¹)	2000	12	377
13. Collecting tank	2000	12	372
14. Dosing pumps	3000	12	558
15. Mixing unit	3000	12	372
Total	97500		34857

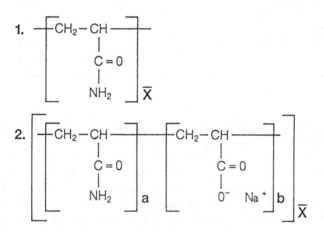

Fig. 10.64. Chemical structure of non-ionogenic flocculants made from technically pure polyacryalmide (1). The anionic flocculants (2) consist of co-polymers of acryalamide (a) with increasing concentrations of acrylate, which creates the negative charge of the polymer. Cationic flocculants are co-polymers synthesized from acrylamide with increasing concentrations of a cationic co-monomer, which ensures the positive charge of the polymer in aqueous solutions.

Table 10.11. *Different flocculants and their impact on harvesting costs*

Flocculant	Amount (p.p.m.)	Price (US$ kg^{-1})	Costs m^{-3} (US$)	Costs kg^{-1} biomass (US$)	Effectiveness
$Al_2(SO_3)_4$	400	1.80	0.70	1.80	+
Sachtoklar	200	0.80	0.15	0.40	+
Sachtoklar	400	0.80	0.30	0.80	+ +
Sachtoklar	900	0.80	0.70	1.80	+ + +
$Al_2(SO_3)_4$	200	1.80	0.35	0.90	
plus Praestol	20	5.80	0.10	0.25	+ + +
Total			0.45	1.15	
$Al_2(SO_3)_4$	150	1.80	0.25	0.65	
plus Praestol	20	5.80	0.10	0.25	+ + +
Total			0.35	0.90	
$Al_2(SO_3)_4$	100	1.80	0.15	0.45	
plus Praestol	15	5.80	0.10	0.20	+ + +
Total			0.25	0.65	
$Al_2(SO_3)_4$	50	1.80	0.10	0.20	
plus Praestol	15	5.80	0.10	0.20	+ +
Total			0.20	0.40	
$Al_2(SO_3)_4$	100	1.80	0.15	0.45	
plus Praestol	10	5.80	0.05	0.15	+ +
Total			0.20	0.60	
Chitosan	200	10.00	2.00	5.00	+
Chitosan	400	10.00	4.00	10.00	+ +

Notes:
Sachtoklar: Polyalumchloride.
Praestol: Anionic flocculant permitted for drinking-water processes.
Chitosan: Hydrolisate of chitinous waste products (i.e. sea food).

The algae suspension falls in such a way as to provide such good soft turbulence and aeration in the flotation tank that the biomass flotates at once without additional aid. Therefore, it has to be investigated whether flocculation and flotation can be done batchwise with several specially constructed harvesting tanks. For the present comparison such equipment is included in the estimation, so that the total harvesting costs by means of flocculation and flotation amount to US$ 1.39 kg^{-1}.

The problem in practice is to find an adequate stirring system to ensure homogeneous distribution of the flocculant. An in-line stirrer, also called a static mixer, can be recommended to optimize stirring. It should be installed between the first stirring unit to which aluminum sulfate will be added and the flotation unit. A flotator works satisfactorily only if a good air decompression unit is installed. Electro-flotators in former use had too many disadvantages due to high electrode wear and are not recommended. Today, new systems are available that produce extremely fine bubbles of a few micrometers in diameter. A model that has been tested successfully in air flotation is the so-called Aquatector. In this unit, water is pumped or supplied under suitable pressure in the device where it passes through fine expansion nozzles creating jet streams of high velocities in an intensive turbulent flow. The air is injected into the mixing chamber and results in supersaturation of the water in a closed system leaving no part of the air going to waste. Undissolved gas is recirculated continuously until fully dissolved. The water leaves the system through a needle-type control valve. The working pressure is comparatively low with 2–5 bar (30–70 psi) only. The microbubbles so produced have an average size of 50 μm. The ascending velocity of gas bubbles in water can be calculated by the equation $V_s = 27.5\ d^{1.4}$, where V_s = the velocity (cm s^{-1}) and d = diameter of bubble (mm). That means that the ascending of such fine 50 μm bubbles is about 0.42 cm s^{-1}.

If it comes to a decision between centrifugation and flocculation, other points besides costs have to be considered. A centrifuge is easy to run and always provides the same product quality. Its efficiency will not be influenced by differences in the culture media, changes in weather, failure of CO_2 supply or other factors that affect the physico-chemical properties of the algae and especially their surface. However, it should be mentioned that a centrifuge, especially when running with a high throughput, does not clarify 100%. In addition, most of the algae are to a certain degree broken and ruptured when desludged, so that any subsequent processing steps must follow immediately to avoid spoiling of the biomass.

Unlike centrifugation, flocculation can be performed anywhere, the algal cells are not destroyed and therefore can be stored longer before processing. It may also be possible to drain the flotate on simple cloths or by means of a decanter. If run properly, the efficiency of flocculation is higher than centrifugation as the supernatant contains less biomass. The major disadvantage of flocculation/flotation is that this method is more complicated to handle. The fluctuating quality of the algal culture owing to exterior influences causes several problems and requires experience and intuition of the personnel running the unit. Jar tests are frequently necessary to find optimum flocculant concentration and combination. Another shortcoming of this harvesting method is the chemical contamination of the concentrated biomass as well as of the filtrate. For instance, the use of 150 p.p.m. flocculants leads to a chemical portion of 35% of the total if the culture has a dry weight concentration of 400 mg l^{-1}. It has not yet been confirmed whether washing of the flocculated biomass with water or dilute HCl removes aluminum sulfate.

Finally, another point, which is often neglected, needs special attention, i.e. the remaining amount of flocculants in that portion of the medium that is recycled into the culture after separation of the algae. Besides aluminum and ferruginous compounds, all other flocculants are organic substances and often composed of polysaccharides, which are easily degradable by microorganisms. In large algal plants with repeated harvesting steps of the same culture, considerable amounts of the flocculants will be accumulated in the medium, thus providing a suitable substrate for the growth of bacteria and other undesirable heterotrophic microorganisms in the initially inorganic medium. This hazard can be minimized, if not avoided, by frequent replacement of at least part of the medium. However, this procedure represents an additional cost factor because the water and all nutrients have to be replaced. Based on the above calculations, the difference between the two harvesting systems is about US$ 0.30. This already relatively small margin has to be judged in view of the several shortcomings and uncertainties mentioned above still connected with the application of flocculants. With regard to all these aspects, it has to be accepted that harvesting algae by centrifugation is at present the most reliable method.

10.8. Drying

In almost all applications where the harvested and concentrated algal biomass is to be utilized further, a product with a water content of less than 10% is required. Moisture affects spoilage of the dried algal product in the sense that it maintains, within a certain range, the growth of bacteria, mould and fungi. In this context one must distinguish between water bound to the biological material in the form of a monomolecular layer and that is not freely available to maintain biological processes (< 2% moisture), and moisture contents above this level at which the water becomes available for microorganisms. A moisture content of 7%, which technically is realistic for most algae production units, is the level that still protects the algae from prolonged microbial growth.

In addition, the algal slurry must be quickly processed, otherwise it will spoil very soon, especially in a hot climate. The only exception probably is the use of a wet algal slurry in direct feeding of livestock or in aquaculture. Therefore, drying is one of the most important steps in algal production and can account for up to 30% of the total production costs. At present, processing is a major economic limitation to the production of low cost algal products and a considerable factor even in the case of higher value products. The various methods that are available for drying concentrated algal slurry are summarized in Table 10.12.

Selection of the particular drying method will depend on the specific algal species, the scale of operation and the final use of the dried product. In

Table 10.12. *General evaluation of available algae drying methods*

Method	Advantage	Limitations	Algae	Remarks
Drum-drying	Fast and efficient	Cost intensive	All algae except *Spirulina*	Ruptures cellulosic cell walls, sterilizes the product, not suitable for *Spirulina*
Spray-drying	Fast and efficient	Cost intensive	All algae	Sterilizes the product, breakage of cellulosic cell walls not always guaranteed
Sun-drying	Very low fixed capital and no running costs	Slow process, weather dependent	All algae	Biomass may ferment, sterilization not possible, does not break cellulosic cell walls
Solar-drying	Low capital costs	Weather dependent	All algae	Does not break cellulosic cell walls, sterilization not possible
Cross-flow-drying	Faster than sun- and solar-drying, cheaper than drum-drying	Requires electricity	All algae	Does not break cellulosic cell walls, sterilization not possible
Vacuum-shelf-drying	Gentle process	Cost intensive	All algae	Does not break cellulosic cell walls, product becomes hygroscopic, sterilization not possible, preserves cell constituents
Freeze-drying	Gentle process	Slow process, cost intensive	All algae	Does not break cellulosic cell walls, sterilization not possible, preserves cell constituents

addition, the various methods differ in the extents of both investment and energy requirement.

Sun-drying

Sun-drying is an easy and least-expensive possibility of drying; however, the method is weather dependent and involves the possible risk of spoilage. At present, the method is restricted to the drying of *Spirulina* because it fails to rupture the cellulosic algae cell walls, which is necessary if the algae are used as food or feed in view of the digestibility of the biomass. Sun-drying is used in Burma and by the *Spirulina*-harvesting population along Lake Chad and other African lakes. It is also relevant in the context of the low-level algal cultivation technology that is being considered in several developing countries.

A common practice for sun-drying of *Spirulina* is to spread the concentrated algae slurry on trays lined with plastic sheets and to expose them in a dust-protected place to the sun (Fig. 10.65). The plastic sheets allow easy removal of the dry algal flakes. Trials using fine-weave fabric instead of the plastic sheets were less effective. Although the drying process could be enhanced by additional air contact with the lower side of the spread algal slurry, the removal of the dried algal flakes from the fabric was more difficult. Furthermore the fabric had to be cleaned very carefully of remaining algal particles that are clogging the pores of the material. Systematic studies on the effectiveness of sun-drying showed that if the thickness of the layer to be dried is greater than 0.75 cm, then it will require more than one day to dry; initial depths of more than 1.3 cm resulted in putrefaction of the algae during the second and third days of exposure. In practice, single-day sun-drying on flat plates was found to be possible with a slurry thickness not exceeding 0.5 cm. At this thickness, an initial concentration of 6% solids and a radiation of $2 \text{ kJ cm}^{-2} \text{ d}^{-1}$, about $130 \text{ g m}^{-2} \text{ d}^{-1}$ of dry algal material can be produced with a moisture content of less than 10%. The method shown is restricted mostly to small cultivation units. On a large scale, sun-drying would entail significant costs for setting up large drying beds lined with plastic and facilities to spread out the algal slurry as a thin film and to collect the flaked dried algae.

It has to be stressed that the method of sun-drying, because of the relatively low temperatures achieved, does not pasteurize or sterilize the dry algal material, which is an important fact to be observed in connection with the utilization of algae for human and, to a lesser extent, animal nutrition.

Solar heater

Solar heaters are a cheap means of shortening the drying time compared with direct sun-drying, and several variations of the basic concept have

Fig. 10.65. Sun-drying of *Spirulina* sp. on aluminum trays lined with plastic sheets.

been described. A simple solar drier, consisting of a wooden chamber with the inside surface painted black and the top covered with a 2 mm glass plate, develops an average temperature of 60–65 °C. A more sophisticated drier, using PVC solar collectors, can further increase the temperature reached within the drying chamber. Such collectors consist of three layers, the first two of which are transparent and the third black: the inner two layers form the collectors whereas the top layer acts as insulation. Air is blown through the two inner layers. Air temperatures as high as 70–75 °C were recorded with a collector 30 m long at a solar radiation of 400 W m^{-2}, an ambient temperature of 31 °C and an air flow rate of 0.3 m s^{-1}.

A cheaper and simpler but quite efficient design is shown in Figs 10.66

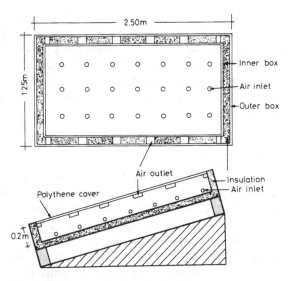

Fig. 10.66. Schematic design of solar drier.

Fig. 10.67. Solar dryer for drying *Spirulina* sp. at MCRC, Madras, India.

and 10.67, having a dimension of about 2.5 × 1.25 m and a depth of 20 cm. It consists of an inner and an outer box, the space between both (5–10 cm) packed with insulation material (grass, fibers, saw dust, rubber foam, etc). The cover is made of two layers of polythene film stretched over the frame, the inside is painted black. Drying is achieved by air flow in the drier induced by air inlet holes directed through the bottom of the boxes. A constant flow of air causes the moisture in the algal material to be evaporated. The dryer can be converted to an all-weather drier by including heating. The additional installation required is a fire box made from clay or bricks and a long metal tubular flow, which acts as a chimney.

In order to increase the efficiency of solar driers, they should be mounted in a sloping position, the degree of inclination being the degree of latitude of the particular location plus 12.

Drum-drying

Although drum-drying requires high costs for investment and energy consumption, it is a reliable alternative for producing a fully digestible and bacteriologically safe product. The principle of this drying method, which is quite common in the food industry, is the application of a wet slurry or paste onto a rotating, preferably chromium-plated, heated drum. The material to be dried is heated for a few seconds and the ensuing dehydration causes the cell wall to open. Depending on the method by which the slurry is fed to the device, two types of drum drier can be distinguished. With the first type, the material is sprayed through fine nozzles onto a single rotating drum, whereas with the second type the material is fed between a small gap formed between two counter-rotating drums ('sump feeding').

Shortcomings of the first system are problems in the uniform distribution of the slurry on the surface of the drum and clogging of the spraying device. The difficulties encountered with the second type are linked with the balanced feeding of the material. If the feeding rate is too high the material will start boiling before it is attached to the drum and dried, thus leading to partial heat-destruction of algal constituents ('Maillard reaction').

The major disadvantage of drum-drying is that the process requires large amounts of energy at a relatively low thermal efficiency. The average evaporating capacity of a drum drier is about $20 \, l \, h^{-1} \, m^{-2}$ of drum surface. Further drawbacks are the facts that highly polished drums are necessary – otherwise the algae tend to stick on the drum surface – and that the knives that remove the dried algae from the drum often become blunt owing to small sand particles present in algal slurry and hence need frequent sharpening. The average costs for drying are estimated at US$ 0.4–$0.6 \, kg^{-1}$ dry material.

Nutritional studies have shown that drum-drying yields an excellent, highly digestible algae powder. The method seems to be less suitable for *Spirulina*, because it may result in an algal product that is slightly burnt, having a very dark green or olive green color and a slightly bitter taste.

Spray-drying

At present the most common drying process for all algae species is spray-drying. Its efficiency is comparable with drum-drying but requires a less concentrated paste (it must be pumpable) than drum-drying and gives a very fine and uniform powder. It has been reported that the product obtained can be less digestible than drum-dried algae, which could be because of the temperatures reached within the sprayed algal slurry droplets being too low and a drying time that is too short. Furthermore, spray-drying may cause significant deteriorations of some algal components such as pigments or vitamins, which may be protected by the addition of antioxidants prior to drying. The cost of this process is comparable with drum-drying (between US$ 0.5 and 1.0 kg^{-1}).

Other methods

There are still other possibilities for drying the concentrated algal slurry. However, these methods are limited to laboratory trials or to the drying of small quantities and are not suitable for large-scale algal production units, mainly because they are too slow and/or too costly. One of these methods is freeze-drying, which is a very gentle process for preserving algae, because it leaves almost all the constituents in their original composition. It produces an attractive, light-green algal powder but does not break the algal cell walls. A similar process is vacuum-shelf-drying, which operates at temperatures between 50 and 65 °C and at about 0.05 atm pressure.

Cross-flow air-drying, which actually represents an improved, artificially heated solar drier, has been tested at various places but has not reached any significant importance as a drying process. It too does not break the algal cell wall and hence is only suitable for drying *Spirulina*.

11 Yield

Among the various aspects connected with algae mass production, there is hardly one that is subjected to more speculation and exaggeration than the yield. A short glance at the respective literatures reveals that data on growth rates of algae vary over a wide range, depending on the species used, the prevailing culture and climatic conditions and, at times, on the particular laboratory or institution; they are sometimes contradictory, depending upon the effect desired by the investigator. Various practical men, and probably still more theorists, have published results of yields obtained or complicated mathematical computations and treatises on yields 'theoretically achievable under optimum conditions'. There are publications that predict unbelievable 'maximum yields of 135 to 346 g organic weight m^{-2} d^{-1}' (Pirt *et al.*, 1980; Raven, 1988). However, colleagues obtaining only one-tenth of these figures should not be disappointed or frustrated because they are very much in the normal range of average algal yields common in large scale production units.

Furthermore, it is misleading and not correct to extrapolate results on growth rates gained by limited studies in the laboratory or on pilot-plant scale to outdoor mass cultivation. One also has to distinguish between results obtained for a very short growth period under optimal conditions and average values determined over a long period. Though algae are one of the best sources, compared with agricultural crops, in terms of protein productivity, the yield predictions have to be made with caution so as to avoid future disappointments.

The yield of algae biomass is expressed normally either as grams (dry matter) produced per square meter (culture area) per day, or more impressively as tons per hectare per year. Sometimes the term grams per liter (g l^{-1}) is found in the literature; however, this dimension does not exactly describe the yield, but simply a concentration, and neither does it reflect the actual situation in algal mass cultivation, where it is not the volume (depth) of the algae pond but the surface (light-receiving) area that is the important factor for algae production.

Most of the data on yields of algae mass cultures reported in the literature represent the increment of algae biomass in the pond and not the actual yield of algae powder. It should be taken into consideration that the computed growth rate does not equal the harvested amount. The differences are normally caused by losses during harvesting operations, standstill periods, infections, etc., so that it is certainly not exaggerated to assume a loss of at least 10% of the gross yield during normal algal plant management.

Several figures are available on growth rates or yields of various algae (Table 11.1). For chlorophyceae they are in the range of 10–35 g m^{-2} d^{-1}, indicating that with the algae production technology of today the optimum yields have probably been reached. Even if it is possible to increase the average yields here and there, it seems that an upper limit of 30–40 g m^{-2} d^{-1} is a given characteristic of the photosynthetic process of algal growth, which cannot be improved further (Goldman, 1979).

For the cyanobactrium *Spirulina*, being a filamentous form with longer doubling time, lower yields of 8–15 g m^{-2} d^{-1} are obtained. The figures given above are the realistic values that must be considered in estimating the feasibility of algae mass production. For sewage systems, comparable data are reported, although the biomass harvested is not specified as algal biomass in several cases, because bacterial biomass is also included.

Several studies on algae production technology that have been conducted during the past four decades and which represent a great variety of culture systems, algal species, geographical locations and pond sizes, have demonstrated a continuous increase in algae productivity, both for short-term maximum and longer-term average yields. This progress becomes evident if we compare the yields of less than 5 g m^{-2} d^{-1} reported in the early 1950s with the values obtained nowadays.

The key factor determining the yields of microalgal systems is irradiance, i.e. light energy input per unit surface and time. In principle, algal suspensions offer the advantage that they can continuously be kept at areal densities that guarantee optimal utilization of photosynthetic active radiation. This requires the exact adjustment of areal density in terms of chlorophyll (see below) per unit of irradiated surface.

By comparing growth rates obtained in laboratory experiments and under field conditions, it becomes evident that the figures of outdoor productivity are considerably lower, an effect commonly attributed to the prevailing light conditions.

Whereas most of the laboratory studies are carried out under low light intensities that are below the photosynthetic saturation point, the outdoor conditions are characterized by high intensities of sunlight. Here the limiting factor is not the maximum total light (quantum) efficiency but the fraction of absorbed radiation that is actually used in photosynthesis. If the algae have a high pigment content, i.e. high light extinction coefficient, but on the other hand perform photosynthesis that saturates at low light intensities, more light will be absorbed than can be utilized in photosynthesis. Depending on the actual saturating and incident light intensities, the extinction coefficients and the culture density, the maximum efficiency will be reduced by a factor of about two to five from that possible at low light

Table 11.1. *Selected data on yields of different algae grown under outdoor conditions*

Algae	Yield ($g\ m^{-2}\ d^{-1}$)	Growth period	Location	Reference
Chlorella pyrenoidosa	18	Summer	Trebon, CSFR	Zahradnik (1968)
Chlorella ellipsoidea	18	May	Trebon, CSFR	Zahradnik (1968)
Chlorella sp.	20	1 year (mixotroph)	Taiwan	Tsukada *et al.* (1977)
Chlorella sp.	15	1 year (autotroph)	Taiwan	Tsukada *et al.* (1977)
Chlorella sp.	21	1 year	Japan	Tsukada *et al.* (1977)
Chlorella sp.	12	1 year	Japan	Tamiya (1957)
Chlorella sp.	17	1 month	Australia	Dodd & Anderson (1977)
Chlamydomonas	7	?	Trebon, CSFR	Nekas & Lhotsky (1967)
Tetraselmis sp.	18	?	Lamezia, Italy	Benemann (1986)
Dunaliella salina	60	'short period'	Israel	Ben-Amotz (1980)
Ankistrodesmus	20	?	Waterfontain, SA	Benemann (1986)
Scenedesmus obliquus	11	240 days	Dortmund, FRG	Stengel & Soeder (1975)
Scenedesmus obliquus	15	1 year	Bangkok, Thailand	Payer *et al.* (1978b)
Scenedesmus obliquus	20	1 year	Mysore, India	Becker & Venkataraman (1982)
Scenedesmus obliquus	30	325 days	Sausal, Peru	Heussler (1985)
Scenedesmus obliquus	15	330 days	Cairo, Egypt	El-Fouly *et al.* (1985)
Scenedesmus obliquus	19	200 days	Rupite, Bulgaria	Dilov *et al.* (1986)
Scenedesmus sp.	19	10 days	Trebon, CSFR	Benemann (1986)
Spirulina sp.	8–12	200 days	Mysore, Thailand	Becker & Venkataraman (1984)
Spirulina sp.	15	?	Bangkok, Thailand	Goldman (1979)
Spirulina sp.	10	1 year	Mexico	Durand-Chastel (1980)

intensities (Benemann, 1989). To overcome this growth-limiting effect, the light intensity that saturates the photosynthesis has to be increased by decreasing the amount of pigments in the algal cell (i.e. the extinction coefficient). If this can be achieved, another problem may arise, namely that superoptimal areal algal densities result in reduced light intensity available for each cell and self-shading effects, which leave a statistical fraction of the algal population below the compensation point resulting in respiratory losses that would reduce the overall yield so that frequent dilution (harvesting) becomes necessary. This, however, is in contrast to the concept that in mass cultures cell densities must be high in order to absorb all the light, which means that the growth rate has to be slow. To overcome this controversy, algal strains that provide maximum rates of photosynthesis at both high and low light intensities have to be found or bred by genetic engineering. This attempt may lead to yet another problem, i.e. that other contaminating alga with a higher content of pigments would overgrow the culture because of its ability to absorb more light in light-limiting zones of the pond. Whether this aim can ever be achieved remains to be seen.

At present it is very questionable that the yields of algae mass cultures can be improved by any method, be it ever so sophisticated, because too many unpredictable and uncontrollable factors are involved.

Another point should be mentioned here, which was observed during cultivation trials with *Scenedesmus*. It is known that changes in external factors influence the growth capacity of algal cells as well as their chemical composition. Light is one of these factors that can change the metabolic behavior of algae.

The geographic location of several algal projects in tropical areas ensures day and night periods that do not fluctuate considerably during the year, i.e. the total darkness is always in the range of 10–12 h. These regular light and dark phases result in partial synchronization of the outdoor cultures of unicellular algae (*Chlorella*, *Scenedesmus*). Darkening of the cells for at least 10 h increases their photosynthetic capacity in the next light phase. The beginning of the dark phase stimulates the cells to their highest physiological capacity and if the algae can produce sufficient cell material during this period they are triggered by the beginning of the following light phase to release autospores. These periodic fluctuations of the physiological activity of the algae results in variations of the chemical composition of the algal cells. With *Scenedesmus* it was found that the protein content in cells was always highest in the morning and lowest in the evening, whereas the carbohydrate content was low in the morning and high in the evening hours. The percentage of protein from morning to evening samples showed variations of 10–15% of total biomass. Hence, if algae are produced predominantly as a source of protein, it is advised to performed harvesting (and processing) in the morning.

With regard to yield, it should be the prime goal of algae cultivation to maximize possible photosynthetic efficiencies and resulting yields by forcing the only major uncontrollable growth factor, the light, to be limiting, which means that all the required nutrients for growth must be supplied in excess.

The dependence of yield on temperature is less pronounced, although the productivity at temperatures below 15 °C becomes fairly low. Whereas most of the chlorophyceae even withstand freezing, *Spirulina* does not grow at low temperatures. On the other hand, overheating may also be of practical importance. Irradiated outdoor cultures may heat above ambient temperature, because most of the incident light energy is transformed to heat, especially if high humidity inhibits cooling by evaporation. For most algae, the maximum tolerable temperature is about 35 °C.

12   Chemical composition

General introduction

As with any higher plant, the chemical composition of algae is not an intrinsic constant factor but varies over a wide range. Several environmental factors influence the proportion of the different constituents of the algal biomass. Very extreme variations in composition were reported in response to parameters such as temperature, illumination, pH-value of the medium, mineral nutrients, CO_2 supply, etc. To obtain an algal biomass with a desired composition, the proportion of the different constituents of several algae can be modified very specifically by varying the culture conditions, for instance nitrogen or phosphorous depletion in the medium or changes of physical factors such as osmotic pressure, radiation intensity, population density, light or dark growth, etc. A wide spectrum of analyses on algal constituents has been published in the literature; a compilation of data on gross chemical composition of different algae is given in Table 12.1 and compared with the composition of selected conventional foodstuffs.

Spoehr & Milner (1949) were probably among the first who published detailed information on the effects of environmental conditions on algal composition and described the effect of varying nitrogen supply on the lipid and chlorophyll content of *Chlorella* and some diatoms. Other authors reported of changes in the composition pattern and amino acid profile of basic proteins.

Variations of carbohydrates, lipids, fatty acids and soluble proteins during the growth cycle of algae were described using the techniques of synchronous culture.

Protein

The high protein content of several microalgal species was one of the main reasons for considering these organisms as an unconventional source of protein. Most of the figures found in the literature on concentrations of algal proteins are based on the estimation of so-called 'crude protein'. As a rule, this figure is obtained by acid hydrolosis of the algal biomass, followed by the estimation of the total nitrogen in the hydrolisate by the method after Kjeldahl (Hiller, Plazin & van Slyke, 1948) and subsequent multiplication of the value by the factor 6.25. However, it has to be kept in mind that this calculation involves certain errors. Because substantial parts of the total

177

Table 12.1 *Gross chemical composition of human food sources and different algae (% of dry matter)*

Commodity	Protein	Carbohydrates	Lipids	Nucleic acid
Baker's yeast	39	38	1	—
Rice	8	77	2	—
Egg	47	4	41	—
Milk	26	38	28	—
Meat muscle	43	1	34	—
Soya	37	30	20	—
Scenedesmus obliquus	50–56	10–17	12–14	3–6
Scenedesmus quadricauda	47	—	1.9	—
Scenedesmus dimorphus	8–18	21–52	16–40	—
Chlamydomonas rheinhardii	48	17	21	—
Chlorella vulgaris	51–58	12–17	14–22	4–5
Chlorella pyrenoidosa	57	26	2	—
Spirogyra sp.	6–20	33–64	11–21	—
Dunaliella bioculata	49	4	8	—
Dunaliella salina	57	32	6	—
Euglena gracilis	39–61	14–18	14–20	—
Prymnesium parvum	28–45	25–33	22–38	1–2
Tetraselmis maculata	52	15	3	—
Porphyridium cruentum	28–39	40–57	9–14	—
Spirulina platensis	46–63	8–14	4–9	2–5
Spirulina maxima	60–71	13–16	6–7	3–4.5
Synechococcus sp.	63	15	11	5
Anabaena cylindrica	43–56	25–30	4–7	—

nitrogen of algae, as well as of other forms of single cell protein (SCP), consist of non-protein nitrogen – which preponderantly originates from nucleic acids, amides, glucosamides or cell-wall materials – the multiplication of total nitrogen by 6.25 results in an overestimation of the true protein content. To avoid this error, purine nitrogen should be determined separately and the amount of nucleic acids computed by special methods. Because the nitrogen content of pyrimidines is about 40% that of purines and both are present in equimolar amounts in most nucleic acids, the purine nitrogen has to be multiplied by 1.4 to obtain nucleic acid nitrogen. Thus, a more correct value for nitrogen is obtained by subtracting 1.4 times the purine nitrogen from crude protein nitrogen.

Other protein estimation procedures, based on color reactions with defined protein constituents and which do not react with other nitrogen-containing compounds, are the method after Lowry (Lowry *et al.*, 1951) and the biuret method (Gornall *et al.*, 1949).

A non-protein nitrogen content of approximately 12% of the total nitrogen was found in *Scenedesmus* sp.; for *Dunaliella* and *Spirulina* values of 6% and 11.5% were reported. As a rule it can be assumed that on average about 10% of the nitrogen found in microalgae consists of non-protein nitrogen.

Amino acids

The nutritional quality of a protein is determined by its amino-acid profile, i.e. content, proportion and availability of the amino acids. Whereas plants are capable of synthesizing all amino acids, animals and humans are limited to the biosynthesis of certain amino acids only (non-essential amino acids); the remaining (essential) amino acids have to be provided through food. For humans, these essential amino acids are leucine, isoleucine, lysine, methionine, phenylalanine, threonine, tryptophan and valine; arginine and histidine are considered semi-essential, because in general they can be synthesized and only during growth or upon the appearance of signs of deficiency must they be provided exogenously. The amino acids cystine and tyrosine can be formed from methionine and phenylalanine if the provision through food is not sufficient.

Because microalgae are considered primarily as a source of protein, the amino-acid composition of the different species in question is of special importance. Table 12.2 shows the amino-acid profile of various algae as compared with egg, soya and a reference pattern of a well-balanced protein, recommended by WHO/FAO. It can be seen that the amino-acid analyses of almost all algae compare favorably with that of the reference and the other proteins. Marginal deficiencies can be seen among the sulfur-containing amino acids methionine and cystein, an observation that is characteristic for many plant proteins.

Lipids

Lipids and fatty acids are constituents of all plant cells, where they function as membrane components, as storage products, as metabolites, and as sources of energy. The basic pathways for lipid synthesis in algae are analogous to those found in higher-plant oil seeds; however, there are certain algal strains that produce unique long-chain or highly unsaturated fatty acids.

The major differences to higher plants are 1) the fact that algae respond to fluctuations of the growth conditions or several types of stress with variations in the fatty-acid composition of the oils; and 2) that in contrast to higher plant cells, the oil-accumulating algal cells are performing photosynthesis, that means that the complete pathway from CO_2-fixation up to triacylglycerol synthesis can be adjusted by the cell. As under all these stress conditions lipid synthesis occurs as the photosynthetic activity decreases, maximum acylglycerol production from oleaginous algae would be obtained in two steps: 1) fast growth under optimum conditions followed

Table 12.2. *Amino-acid pattern of different algae as compared with conventional protein sources and a reference pattern (g per 16 g N)*

Amino acid	Source									
	1	2	3	4	5	6	7	8	9	10
Ile	4.0	6.6	5.3	6.0	6.7	3.6	4.5	3.4	3.2	4.2
Leu	7.0	8.8	7.7	8.0	9.8	7.3	9.3	4.0	9.5	11.0
Val	5.0	7.2	5.3	6.5	7.1	6.0	7.9	5.1	7.0	5.8
Lys	5.5	5.3	6.4	4.6	4.8	5.6	5.9	7.9	6.4	7.0
Phe	—	5.8	5.0	4.9	5.3	4.8	4.2	4.5	5.5	5.8
Tyr	6.0	4.2	3.7	3.9	5.3	3.2	1.7	2.7	2.8	3.7
Met	—	3.2	1.3	1.4	2.5	1.5	0.6	1.8	1.3	2.3
Cys	3.5	2.3	1.9	0.4	0.9	0.6	0.7	—		1.2
Try	1.0	1.7	1.4	1.4	0.3	0.3		1.4	—	0.7
Thr	4.0	5.0	4.0	4.6	6.2	5.1	4.9	3.2	5.3	5.4
Ala	—	—	5.0	6.8	9.5	9.0	12.2	5.9	9.4	7.3
Arg	—	6.2	7.4	6.5	7.3	7.1	5.8	5.6	6.9	7.3
Asp	—	11.0	1.3	8.6	11.8	8.4	8.8	5.9	9.3	10.4
Glu	—	12.6	19.0	12.6	10.3	10.7	10.5	9.3	13.7	12.7
Gly	—	4.2	4.5	4.8	5.7	7.1	10.4	4.8	6.3	5.5
His	—	2.4	2.6	1.8	2.2	2.1	1.7	1.4	2.0	1.8
Pro	—	4.2	5.3	3.9	4.2	3.9	5.0	4.0	5.0	3.3
Ser	—	6.9	5.8	4.2	5.1	3.8	5.2	2.2	5.8	4.6

Notes:
1, reference protein (FAO/WHO, 1973); 2, egg; 3, soya (Diem & Letner, 1975); 4, *Spirulina maxima* (Clement, Giddey & Menzi, 1976b); 5, *Spirulina maxima* (Becker & Venkataraman, 1984); 6, *Scenedesmus obliquus* (Becker, 1984); 7, *Chlorella ellipsiodae* (Priestley, 1976); 8, *Chlorella pyrenoidosa* (Lubitz, 1963); 9, *Chlorella vulgaris* (El-Fouly *et al.*, 1985); 10, *Dunaliella bardawil* (Ben-Amotz & Avron, 1980).

by 2) imposition of nitrogen starvation or other stress factors with the aim of achieving maximum lipid content. Furthermore, many renewable algae systems might become competitive if petroleum requires more energy for exploration and delivery than it will yield, and although most of the oils produced by microalgae have fatty acid constitutions similar to common vegetable oils, there are several high-value unusual fatty acids that can be synthesized exclusively by a variety of algae.

Lipids extracted with lipophilic organic solvents (ether, petroleum ether, chloroform, etc.) are called commonly 'total lipids'.

Lipids can be classified according to their polarity, which depends on the non-polar (lipophilic) carbon chains (fatty acids) and the polar (hydrophilic) moieties (carboxylic groups, alcohols, sugars, etc). The major part of the non-polar lipids (neutral lipids) of microalgae are triglycerides and free

Table 12.3. *Range of lipid levels found in different classes of algae*

Algal class	Total lipids (% of dry matter)	Percentage of total lipids		
		Neutral	Glycolipids	Phospholipids
Cyanobacteria	2–23	11–68	12–41	16–50
Chlorophyceae	1–70	21–66	6–26	17–53
Crysophyceae	12–72	—	—	—
Prymnesiophyceae	5–48	—	—	—
Cryptophyceae	3–17	—	—	—
Xanthophyceae	6–16	44	17	39
Rhodophyceae	1–14	41–58	42–59	—
Bacillariophyceae	1–39	14–60	13–44	10–47

Source: After Borowtzka, 1988.

fatty acids, whereas the polar lipids are essentially glycerides in which one or more of the fatty acids has been replaced with a polar group, for instance phospholipids and glycolipids.

The average lipid content varies between 1 and 40%, and under certain conditions it may be as high as 85% of the dry weight. Algal lipids are typically composed of glycerol, sugars or bases esterified to fatty acids having carbon numbers in the range of C12–C22. Most of the fatty acids found in algal lipids are straight-chain molecules with an even number of carbon atoms as a result of their biosynthesis from acetate by α-addition. They may be either saturated or unsaturated; some cyanobacteria, especially the filamentous species, tend to have larger amounts of polyunsaturated fatty acids (25–60% of the total). On the other hand, those species that show facultative anoxygenic CO_2 photoassimilation with sulfite as electron donor lack polyunsaturated fatty acids.

Eukaryotic algae predominantly contain saturated and monosaturated fatty acids; triglycerides are the most common storage lipids constituting up to 80% of the total lipids fraction. Other major algal lipids are sulfoquinovosyl diglyceride, mono- and digalactosyl diglyceride, lecithin, phosphatidyl-glycerol and γ-inositol. A summary of the range of lipid contents reported in different algal taxa is given in Table 12.3.

Nutritional and environmental factors can affect both the relative proportions of fatty acids as well as the total amount. Many microalgae growing under nitrogen limitation show enhanced lipid contents. In the late 1940s it was noted that nitrogen starvation is most influential on lipid storage and lipid fractions as high as 70–85% of dry weight were reported. Some algal species such as *Dunaliella* sp. or *Tetraselmis suecica*, however, respond with decreasing lipid contents and produce carbohydrates rather than lipids under such conditions. Besides nitrogen, other nutrient deficiencies may also lead to an increase in the lipid content. In diatoms, for instance, the amount of lipids can be increased during silicon starvation. The effect of different nitrogen regimes on lipid content and other growth parameter of different algae (*Scenedesmus obliquus, Chlorella vulgaris, Anacystis nidulans, Microcystis aeruginosa, Oscillatoria rubescens* and *Spirulina platensis*) was studied by Piorreck, Baasch & Pohl (1984). The authors found that at low nitrogen levels chlorophyceae contained high percentages of total lipids (45% of the biomass) and that 70% of these were neutral lipids such as triacylglycerols (containing mainly 16:0 and 18:1 fatty acids). At high nitrogen levels, the percentage of total lipids dropped to about 20%; in this case the predominant lipids were polar lipids containing polyunsaturated C16 and C18 fatty acids. More specifically, during the initial stages of growth, the algae produced larger amounts of polar lipids and polyunsaturated fatty acids, whereas towards the end of the growth the main lipids were neutral with mainly saturated fatty acids. Cyanobacteria do not show any significant changes in their fatty acid and lipid composition at different nitrogen supply, confirming the general assumption that chlorophycea rather than cyanobacteria can be manipulated in mass cultures to yield a biomass with desired lipid composition. *Dunaliella* sp., when grown under optimal conditions, contains predominantly highly unsaturated C16–C18 fatty acids, whereas under nitrogen limitation an increase of 18:1 fatty acids and the synthesis of C20 fatty acids can be observed.

Not much is known about the influence of nitrogen starvation on the lipid metabolism in other algal species. It has been reported that *Porphyridium cruentum* doubles its total lipid content (mainly neutral lipids) under nitrogen starvation. Other factors than nitrogen nutrition have also been noted to influence the lipid production in algae. For instance, a shift in lipids to triglycerides with increasing light intensities has been observed for *Nitzschia closterium*.

Light enhances the formation of polyunsaturated C16 and C18 fatty acids as well as mono- and digalactosyldiglycerides in *Chlorella*. The effect of varying CO_2 supply on the fatty-acid composition has been described for *Chlorella fusca*. Autotrophically grown cells supplied with 1% CO_2 in air showed 16:0, 16:3, 16:4, 18:1, 18:2 and 18:3 as the major fatty acids with a considerable amount of 16:4. In cultures supplied additionally with glucose, the 16:4 acid disappeared and the 18:3 content was reduced, whereas the other acids increased in a complementary manner.

Low temperatures have been reported to favor the synthesis of polyunsaturated C18 fatty acids in some algae and to cause changes in the fatty-acid composition of *Dunaliella*.

Certain variations can be found between the different taxa. In chlorophyceae, as in higher plants, linolenic acid (18:3) is common, whereas in bacillariophyceae palmitic acid (16:0), hexadecenoic acid (16:1), and C20-polyenoic acid are the major fatty acids. Rhodophyceae such as *Porphyri-*

Table 12.4. *Analytical data on fatty-acid composition of lipids of different algae*

Fatty acid	Spirulina platensis	Spirulina maxima	Scenedesmus obliquus	Chlorella vulgaris	Dunaliella bardawil
12:0	0.4	tr	0.3	—	—
14:0	0.7	0.3	0.6	0.9	—
14:1	0.2	0.1	0.1	2.0	—
15:0	tr	tr	—	1.6	—
16:0	45.5	45.1	16.0	20.4	41.7
16:1	9.6	6.8	8.0	5.8	7.3
16:2	1.2	tr	1.0	1.7	—
16:4	—	—	26.0	—	3.7
17:0	0.3	0.2	—	2.5	—
18:0	1.3	1.4	0.3	15.3	2.9
18:1	3.8	1.9	8.0	6.6	8.8
18:2	14.5	14.6	6.0	1.5	15.1
18:3	0.3	0.3	28.0	—	20.5
18:3	21.1	20.3	—	—	—
20:2	—	—	—	1.5	—
20:3	0.4	0.8	—	20.8	—
Others	—	—	2.5	19.6	—
Reference	Hudson & Karis (1974)	Hudson & Karis (1974)	Keynon *et al.* (1972)	Spoehr & Milner (1949)	Fried *et al.* (1982)

Note:
tr, traces.

dium sp. have high levels of arachidonic acid (20:4) as well as palmitic, oleic (18:1) and linoleic acids (18:2). Chrysophytes contain considerable amounts of 22:4 and 22:6 fatty acids, and in dinophyceae larger quantities of highly unsaturated fatty acids with 16, 18, 20 and 22 carbon atoms can be found (Table 12.4).

Among the various fatty acids, the commercially important ones are the essential polyunsaturated fatty acids, namely linoleic acid (18:2 ω9, 12), γ-linolenic acid (18:3 ω6, 9, 12), dihomo-linoleic acid (20:3 ω8, 11, 14), arachidonic acid (20:4 ω5, 8, 11, 14) and eicosapentaenoic acid (20:5 ω5, 8, 11, 14, 17). For instance, γ-linolenic acid is regarded as effective in lowering the plasma cholesterol level and has been used as dietary supplement for the treatment of various diseases. Because this fatty acid is very rare in common foodstuffs, appreciable amounts are extracted from the evening primrose and added as a substitute to various food and dietary preparations. It has been found that *Spirulina platensis* may serve as a valuable source of γ-linolenic acid, as about 20–30% of its fatty acids consist of this compound.

In the same alga, poly-β-hydroxybutyrate has been identified as a lipid reserve that accumulates during exponential growth up to 6% of dry weight. It can be assumed that besides the different *Spirulina* species, other cyanobacteria may also synthesize and accumulate this compound in larger amounts under certain culture conditions.

The red alga *Porphyridium cruentum* is one of the richest natural sources of arachidonic acid, which constitutes about 36% of the total fatty acids at the usual culture temperature of 25 °C. At lower temperatures (16 °C) the amount may increase up to 60%. Although our present knowledge is limited to a small number of species, it is evident that the lipid fraction of the algae is characterized by a great variability, which, to a certain extent, can be controlled and modified by environmental and culture conditions. Cyanobacteria show an intrinsic variability in the fatty-acid composition that only scarcely can be influenced by external factors; chlorophycea, on the other hand, respond markedly to changes of the environment.

As described above, microalgae may contain significant quantities of lipids that resemble vegetable and fish oils, and could therefore be considered as a substitute for petroleum products, obtained by direct extraction and refinement. They can also be used as vegetable-oil substitute, either in human and animal nutrition or as raw product for the production of plasticizers or cosmetics.

Certain algal lipids can be used as surfactants with properties different from those of synthetic compounds. The advantage of the algal-derived surfactants, which are mainly phosphatidylglycerol, phosphatidylcholine and different galactosyl diglycerides, is the fact that they are biodegradable.

Some of the essential unsaturated fatty acids found in algal lipids are of pharmaceutical importance. They are precursors of prostaglandins, prostacyclins and leucotrienes and as such are becoming increasingly important in the pharmaceutical industry. Patents have been applied for their use as antihypertensives, for treatment of hyperlipidemia, for cholesterol reduction, and as health foods, etc.

However, all these very promising sounding applications of algal lipids should always be considered realistically with regard to the current economic situation. By comparing the present world market prices for salmon oil (14% eicosapentaenoic acid, 12.4% docosahexaenoic acid, 2.8% docosapentaenoic acid) of US$ 5, and for fish oil (18% eicosapentaenoic acid and 12% docosahexaenoic acid) of US$ 6, one has to accept that under the given conditions algal lipids can by no means compete with conventional sources of polyunsaturated fatty acids, even by considering a better price for these lipids because they originate from a plant source.

In the future, a few algae species may be utilized as an important renewable source of liquid fuels, which can be produced by various methods. Microalgae biomass production as a source of fuels was first proposed over thirty years ago, subjected to several laboratory and engineering–economic analyses and revived by the recent 'energy crisis'.

One of the possibilities for producing algal fuel is the direct extraction of lipids and their processing as a diesel-fuel substitute. However, most of the algal lipids have a lower fuel value than does diesel fuel and therefore require pretreatment (heating, combustion) for satisfactory performance in engines. In addition, algal lipids, like many other vegetable oils, contain several by-products that may cause problems with engine performance so that purification steps have to be introduced.

A second method for modifying algal lipids is transesterification to ester fuels, which also may serve as a substitute for petroleum-derived diesel fuels, having a fuel value about 10% lower than diesel fuel. This process is accomplished by the addition of excess alcohol and an alkaline catalyst. The conversion efficiency is about 98% with the production of glycerol as by-product.

However, calculations have shown that at an algal productivity of 75 t $ha^{-1} y^{-1}$ and a lipid content of the algal biomass of 30%, the present costs per liter of algal-derived oil would be more than US$ 1, prior to extraction and processing costs, which may add another 20% so that a commercial production is still far away. Hydrogenation of algal biomass at high pressure and temperature for producing hydrocarbons as liquid fuel has been described in the literature, but not much is known about the economics of this process.

Summarizing the available information, it can be stated that microalgae represent a valuable source of a wide spectrum of fats and oils with different application potentials. New developments in the chemical industry will further enhance the range of commercially important products, for instance in the replacement of petrochemical feedstocks.

Waxes

Certain algae contain considerable amounts of waxes, composed of esters of long-chain fatty acids with long-chain alcohols. They are, for instance, accumulated in aerobically grown *Euglena gracilis*, which converts part of its storage polysaccharides to waxes. Levels of up to about 50% of dry weight could be obtained by first growing the algae heterotrophically and then maintaining them anaerobically for a few days. The wax consists of chain lengths between C27 and C30 but can be changed by modifying the carbon source for growth.

When the cryptomonad *Chroomonas salina* is grown photoheterotrophically on sea water containing 0.25 M glycerol, wax esters are formed that may amount to 70% of the total lipids. Up to the present, only very few algal species have been screened for their capability of synthesizing waxes and hence it is quite likely that other euglenophytes or representatives of other algal classes will be found with a potential for wax production.

Hydrocarbons

Although hydrocarbons (C_{15}–C_{27}) have been detected in small amounts (less than 5% of dry matter) in various chlorophyceae and cyanobacteria, there are only a very few species that contain relatively large amounts. To date, only the two species *Dunaliella* sp. – as a source of carotenoids – and *Botryococcus braunii* – as a source of a mixture of unique C_{17}–C_{34} hydrocarbons – have been shown to produce larger amounts, of which the latter has been implicated as a potential producer of renewable liquid fuels.

This latter alga has the capacity to divert part of its photosynthate into several distinct acyclic olefin molecules, synthesized via the terpenoid pathway and summarized under the term botryococcenes. The main molecule named botryococcene has the sum formula $C_{34}H_{58}$.

During exponential growth, *Botryococcus* has a hydrocarbon content of about 20%. Under unfavorable conditions, however, the alga enters a stage where the content of unsaponifiable lipids increases to levels of up to 90%. The chain length of the hydrocarbons, which under optimal growth conditions varies between C_{27} and C_{31}, shifts to mainly C_{35} (botryococcene and *iso*-botryococcene) and consists of mixtures of monoenes, dienes and trienes. These hydrocarbons are produced internally and often are excreted to the outside of the cell where they are accumulated as an oleaginous layer around and between the cells of this colonial organism.

It has been reported that the hydrocarbon mixture can be cracked into the same distillation products as petroleum (Jassby, 1988).

Almost all the information collected on *Botryococcus* has been gathered from algae cultivated indoors; mass cultivation and bulk production of hydrocarbons outside the laboratory have not yet been demonstrated. This is mainly for two reasons. The first is the low doubling time, which is in the order of more than seven days. Because hydrocarbons are an energy-intensive product, the mass accumulation of a hydrocarbon-synthesizing alga will be less than that of the typical unicellular green alga. However, increased productivity seems to be possible by changing the culture conditions and by using certain selected strains, modifications which, according to the literature, resulted in doubling times of two to three days. Hence, it seems worthwhile to examine the different strains for better growth performance and productivity.

Secondly, there is the fact that high hydrocarbon concentrations exist only in non-growing, senescent and even decaying cultures; in the growing stage the amount of hydrocarbon is fairly low.

The lipid fraction of the chlorophycea *Dunaliella salina* amounts to 50% of the organic matter, and large proportions of the total lipids consist of cyclic and acyclic hydrocarbons, the amount of which can be influenced by temperature and light intensity.

Small amounts of hydrocarbons have also been described for various other algae. Several marine and fresh-water species contain $C_{21:6}$ hydrocarbons; in phaeophyceae C_{15} and in rhodophyceae C_{17} components have been analyzed as the predominating hydrocarbons. The hydrocarbons of most cyanobacteria consist of normal alkanes, alkenes and methyl-branched alkanes, mainly with C_{17} components. Some species such as *Anacystic montana* also have some higher molecular weight hydrocarbons in the range of C_{21} to C_{36}.

Glycerol

All plant cells, including microalgae, are surrounded by membranes, which separate the cell cytoplasma from the environment. These membranes are more or less permeable to water but almost impermeable to solutes. In this context, it is essential for all cells to maintain a balance between the osmotic pressure within the cell and its surroundings. Within certain limits the cell is able to respond to osmotic fluctuations of the aqueous environment by water uptake or excretion, driven by the difference of the chemical potential that develops across the cell membrane, thus decreasing or increasing the concentration of solutes within the cell. In response to higher salt concentrations in its surroundings, the cell counterbalances the rise in outside solutes by enhanced synthesis of solutes inside the cytoplasma and/or increased uptake of solutes from outside.

Algae represent a group of plants that exhibit a very wide range of tolerance to different salt concentrations in their environment, ranging from millimolar amounts up to saturated brine. With regard to this ability to withstand low or high salt concentrations, the algae may be divided into halotolerant and halophilic species, the latter requiring salt for optimum growth and the former having osmoregulatory mechanisms that permit survival in saline media.

The most outstanding example of osmoregulation is the halophilic green alga *Dunaliella* with its different strains, which can thrive in media containing 0.1–8 M (saturated) salt solutions. The major organic osmoticum of this alga that enables the cell to withstand the outer osmotic pressure is glycerol, and the extraction and commercialization of this compound has been suggested.

The economic potential of this biological glycerol production was studied in detail, especially in Israel (Ben Amotz & Avron, 1983). It was found that under optimum conditions, *Dunaliella* can accumulate up to 40% of glycerol on a dry weight basis or, if expressed in the common algal production dimension, 16 g of glycerol m^{-2} d^{-1}. Long-term outdoor experiments showed that an average production of about 4.5 g glycerol m^{-2} d^{-1} can be obtained at a salt concentration of 3.5 M. Because the conditions for optimal increase in total biomass are not necessarily those that maximize glycerol production, a compromise between algal growth and glycerol synthesis has to be found. Hence, the highest yield of glycerol will be obtained in about 2 M NaCl, whereas the conditions favoring maximal algal growth are at a lower salt concentration. It has been suggested that the several osmoregulators that can be found in different algae (glycerol, amino acids, carbohydrates, mannitol, mannose, sorbitol, etc.) protect enzyme activity at high osmotic pressure, an assumption supported by the fact that several enzymes are only slightly inhibited by 5 M glycerol while being severely inhibited by much lower concentrations of salt. There are other plans to utilize the glycerol-rich algal biomass itself as a substrate for a variety of bioconversion processes.

However, this idea of glycerol production by means of algae has some drawbacks, because the osmotic activity of glycerol is less than that of sucrose or sorbitol. In addition, equimolar solutions of NaCl and glycerol do not result in similar osmotic potentials. Practice has shown that up to 45% of the actual external NaCl concentration can be found within the algal cell and only the compensation for the external osmotic pressure is made up by glycerol.

Although the idea of biological glycerol production sounds very promising, it still faces several technological problems such as harvesting of the algal cells and extraction of the glycerol. It is a fact at present that the price of glycerol produced this way cannot compete with glycerol produced synthetically.

Vitamins

Microalgae biomass represents a valuable source of nearly all important vitamins, which improve the nutritional property of this unconventional protein. Unfortunately, information on the vitamin content of different algae is rare and scattered in the literature, probably because the determination of various vitamins in a particulate material such as microalgae is rather difficult. As in higher plants, the vitamin content in algae varies depending on the respective growth conditions.

In addition to these fluctuations caused by environmental factors, the methods applied for drying and processing the algae reduce the concentrations of several vitamins. This is true especially for the heat-unstable vitamins B_1, B_2, C, and nicotinic acid, the concentration of which decreases considerably during the drying processes, compared to the fresh material.

A compilation of the amounts of the major vitamins determined in algae, with those in spinach and beef liver, is given in Table 12.5. Besides these, other vitamins (vitamin K, isomers of tocopherol, etc.) and metabolic intermediates can be found in almost all algae. The concentrations of the vitamins are comparable between the different algae as well as other microbial sources such as bacteria and yeasts. Most of the vitamins listed

Table 12.5. *Vitamin content of different algae (mg kg^{-1} dry matter) in comparison with conventional food sources and recommended daily intake*

Vitamin	RDI (mg d^{-1})	Beef liver (fresh)	Spinach (fresh)	1	2	3	4	5	6
Vitamin A	1.7	360	130	840	225	—	230	554	480
Thiamine (Vit. B_1)	1.5	3	0.9	44	14	55	8	11.5	10
Riboflavin (Vit. B_2)	2.0	29	1.8	37	28.5	40	36.6	27	36
Pyridoxine (Vit. B_6)	2.5	7	1.8	3	1.3	3	2.5	—	23
Cobalamine (Vit. B_{12})	0.005	0.65	—	7	0.3	2	0.4	1.1	0.02
Vitamin C	50	310	470	80	103	—	20	396	—
Vitamin E	30	10	—	120	—	190	—	—	—
Nicotinate	18	136	5.5	—	—	118	120	108	240
Biotin	—	1	0.07	0.3	—	0.4	0.2	—	0.15
Folic acid	0.6	2.9	0.7	0.4	—	0.5	0.7	—	—
d-Ca-Pantothenate	8	73	2.8	13	11	11	16.5	46	20

Notes:
RDI, Recommended daily intake (adults).
1, *Spirulina platensis* (Faggi, 1980); 2, *Spirulina maxima* (Jaya, Scarsino & Spadoni, 1980); 3, *Spirulina* sp. (Durand-Chastel, 1980);
4, *Scenedesmus obliquus* (Becker, 1984); 5, *Scenedesmus quadricauda* (Cook, 1962); 5, *Chlorella pyrenoidosa* (Fisher & Burlew, 1953).

have also been found in supernatants of algal cultures, either released actively or by disintegration of the algal cell.

Plant foods are not generally considered a source of vitamin B_{12}, thus its detection in green algae is rather surprising. It is assumed that these algae do not have the ability to synthesize, but to absorb and to concentrate vitamins provided by closely associated bacteria. The true detection of relatively high concentrations of vitamin B_{12} in *Spirulina* indicates a close phylogenetic link of this algae to bacteria, which are able to synthesize this vitamin.

Claims of the distributors of *Spirulina*-containing health food preparations lead one to believe that 1 g of *Spirulina* contains about 0.5–2.0 µg of vitamin B_{12}. However, radioassays, using human-saliva R binder to measure total corrinoids (cobalamins plus all other molecules containing the corrin nucleus) and pure hog intrinsic factor to measure intact pure cobalamins alone, i.e. true vitamin B_{12}, suggest that more than 80% of what appears to be vitamin B_{12} by standard microbiological assay using *Lactobacillus leichmannii* is in fact analogues of B_{12}. These observations agree with earlier reports indicating that several organisms synthesize a variety of vitamin B_{12} analogues that have no vitamin B_{12} activity for humans. In this connection it cannot be excluded that some of the analogues found in *Spirulina* could be vitamin B_{12} antagonists, which even might be harmful to humans.

There is, however, no doubt that *Spirulina* is a true source of vitamin B_{12}, as could be shown in animal-feeding experiments using rats that were depleted for six weeks with a B_{12}-deficient diet, followed by a four-week repletion period, in which the rats were supplemented with equal doses of vitamin B_{12} (pure cyanocobalamin) or dried *Spirulina*. No differences in the major body parameters (weight gain, liver weight, kidney weight, hematology) could be found between either groups during depletion or repletion. However, cobalamin contents of serum and kidney were lower whereas liver cobalamin content was higher for *Spirulina*-fed animals than for the cyanocobalamin-supplemented controls, indicating that cobalamin from algae was absorbed by the rats but that, as evidenced by the different distribution pattern, at least part of the cobalamins may be analogues.

Pigments

Chlorophyll
One of the most obvious and arresting characteristic of the algae is their color. In general, each phylum has its own particular combination of pigments and an individual color. Remembering the phylogenetic age of the algae it is not surprising that they have developed several pigments that are peculiar to them, in addition to those also found in higher plants. With a few exceptions, the pigmentation of higher plants is similar to that of the

chlorophyta, the phylum from which land plants are thought to have evolved.

All algae contain one or more type of chlorophyll; for the nine recognized phyla of algae, five different chlorophylls have been described. Chlorophyll-*a* is the primary photosynthetic pigment in all algae and is the only chlorophyll in the cyanobacteria (blue-green algae) and the rhodophyta. Like all higher plants, chlorophyta and euglenophyta contain chlorophyll-*b* as well; the additional chlorophylls -*c*, -*d*, and -*e* can be found in several marine algae and fresh-water diatoms. Chlorophylls amount usually to about 0.5–1.5% of dry weight.

Carotenoids

The second important group of pigments found in algae are the carotenoids. Carotenoids are yellow, orange or red lipophile pigments of aliphatic or alicyclic structure composed of eight, five-carbon (isoprenoid) units, which are linked so that the methyl-groups nearest the center of the molecule are in the 1,5-positions whereas all other lateral methyl-groups are in the 1,6-positions. Lycopene, synthesized by stepwise desaturation of the first 40-carbon polyene phytoene, is the precursor of all carotenoids found in algae.

The carotenoids can be divided into two main groups, i.e. pigments composed of oxygen-free hydrocarbons, the carotenes, and their oxygenated derivates, the xanthophylls, which contain epoxy-, hydroxy-, ketonic-, carboxylic-, glycosidic-, allenic-, or acetylene-groups. All algae contain carotenoids, each species usually between five and ten major forms, the variety of which in algae is greater than in higher plants.

Although certain carotenoids occur in most algal classes (for example β-carotene, violaxanthin, neoxanthin), other carotenoids are restricted to few algal classes only. The average concentration of carotenoids in algae amounts to about 0.1–2% of the dry weight. In some algae, however, for instance *Dunaliella bardawil*, when cultivated under appropriate growth conditions (high light intensity and high salinity of the medium), the amount of β-carotene can be manipulated to vary from about 2% to about 14%, accumulated as small droplets within the chloroplast. The primary carotenoids are associated with chlorophyll in the thylakoids, where they act to trap light energy and to transfer it to the photosystem. Several pigments have photoprotective functions, protecting the chlorophyll molecules against bleaching and destruction through extreme radiation or oxygen. It is this function as a lipid antioxidant that may be responsible for the possible therapeutic property of β-carotene as an anticancer agent. Some carotenoids undergo light-induced epoxidation and de-epoxidation or play a role in phototropism and phototaxis.

The presence and importance of β-carotene isomers in plants and algae has been a subject of controversy and it was suggested that the detection of isomers may be an analytical artefact, caused by isomerization during long extraction and purification procedures. However, new HPLC analysis methods have provided evidence for the natural occurrence of different xanthophyll and carotenoid isomers in plants and algae.

The similarity between chlorophyta and higher plants is attested by the common carotenoid pattern, consisting of β-carotene, lutein, antheraxanthin, violaxanthin and neoxanthin. Besides these major pigments, small amounts of additional carotenoids have been detected in chlorophyta: α-carotene, loroxanthin, siphonoxanthin, pyrenoxanthin, siphonein, lycopene and cryptoxanthin.

The pigments of *Scenedesmus obliquus*, one of the chlorophyta investigated in most detail during the early studies of algal mass cultivation, have been identified as: β-carotene, traces of α-cryptoxanthin, lutein, small amounts of zeaxanthin, violaxanthin, loroxanthin and neoxanthin. For *Dunaliella*, the second green alga of commercial importance, β-carotene, α-carotene, *cis*-α-carotene, lutein, lutein 5,6–epoxide, antheraxanthin, violaxanthin, zeaxanthin and neoxanthin have been reported (Borowitzka & Borowitzka, 1988).

The pigment composition of the procaryotic cyanobacteria differs from that found in chlorophyta, as their characteristic carotenoids are echinenone and zeaxanthin. In addition, cyanobacteria are able to synthesize carotenoid glycosides; the predominant representatives of this type are myxoxanthophyll, oscillaxanthin and aphanizophyll.

The lipophile pigments of *Spirulina*, the only cyanobacterium of commercial importance so far, have been determined as: β-carotene, echinenone, α-cryptoxanthin, zeaxanthin, two myxol glycosides containing 6-deoxy-L-glucose and 6-deoxy-L-galactose and two oscillol-glycosides containing the same sugar moieties. It was shown that the carbohydrates of the four glycosides are esterfied with fatty acids.

The β-carotene of *Dunaliella bardawil*, when cultivated under defined conditions, is composed of approximately 50% all-*trans* β-carotine and 40% 9-*cis*-β-carotene (see Fig. 12.1). The 9-*cis* to all-*trans* ratio is proportional to the integral light intensity to which the algae are exposed during a division cycle. In cells grown under continuous white light of 2000 μE m^{-2} s^{-1} the ratio is about 1.5, whereas in cells grown under a light intensity of 50 μE m^{-2} s^{-1}, the ratio is around 0.2 (Ben-Amotz, Lers & Avron, 1988). The findings suggest that the isomerization reaction that leads to the formation of the *trans*-isomer occurs early in the path of carotene biosynthesis, at or before the synthesis of all-*trans* phytoene. It has been shown that in the presence of norflurazon, a herbicide that blocks the conversion of phytoene to β-carotene, *D. bardawil* accumulates massive amounts of phytoene in place of β-carotene. The occurrence of β-carotene isomers and the influence of light intensity on its formation is not confined to algae but seems to be common to many higher plants. The effect of light

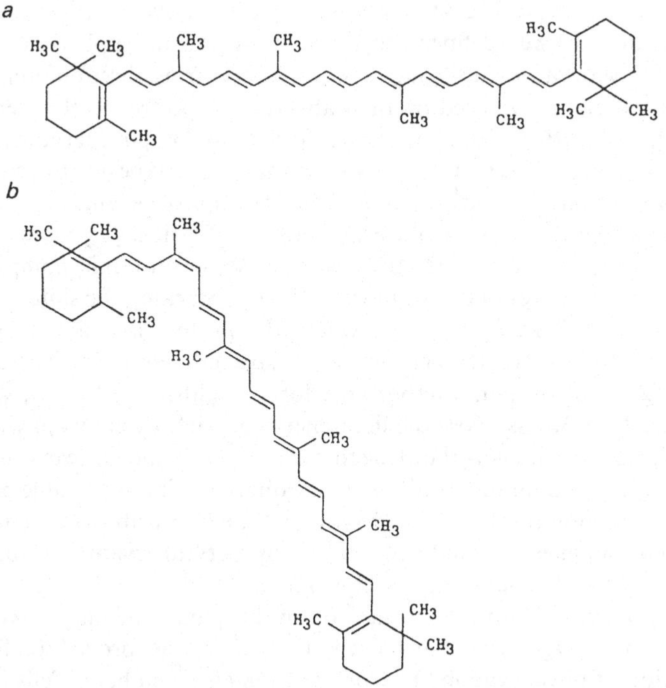

a

b

Fig. 12.1. Structural formula of all-*trans* β-carotene (*a*) and 9-*cis*-β-carotene (*b*).

could be demonstrated with carrots where the parts of the plant exposed to light (leaves and flowers) contained *trans*-and *cis*-isomers whereas in the root only all-*trans*-β-carotene could be found.

Among the numerous carotenoids found in the algae, only a small number are of commercial importance, these include β-carotene, lycopene, zeaxanthin, astaxanthin and lutein, which are used mainly as food colorings or as feed additives. At present, they are sold in small quantities, mainly because of their high cost, and the majority of the commercially used carotenoids are still produced synthetically. However, the increasing number of algal plants, especially those designed for the production of the β-carotenoid-rich *Dunaliella*, may soon change the situation.

Phycobiliproteins

Besides the lipophile pigments, cyanobacteria, rhodophyta and cryptophyceae contain phycobiliproteins, deep-colored water-soluble proteinaceous accessory pigments, which are components of a complex assemblage, the phycobilisomes. The definition for purified phycobiliproteins is that they are proteins made up of α–β monomers of two dissimilar polypeptide chains, α and β, in which the prosthetic group, the bile pigment, is tightly bound by covalent linkages to the apoprotein. The phycobiliproteins can be divided into two well-defined prosthetic groups, forming phycoerythrobilins and phycocyanobilins. Among the phycoerythrins, three different types can be distinguished based on their absorption spectra: R-phycoerythrin and B-phycoerythrin, found in rhodophyceae, and C-phycoerythrin, present in cyanobacteria. Corresponding to that, three types of phycocyanin have been isolated: R-phycocyanin from rhodophyceae and C-phycocyanin and allophycocyanin from cyanobacteria. Purified phycobiliproteins may be monomers, dimers, trimers or hexamers of this complex, or equilibrium mixtures of two or more of these aggregates, the state of which depends on the organism, from which the protein is isolated and the methods applied (pH, temperature, ionic strength, etc.). The existence of these pigments is of major importance for the light-harvesting capabilities of algae. They almost close the light-energy gap left by chlorophyll-*a* and the carotenoids, allowing the algae to use solar radiation more efficiently in a manner like that of fucoxanthin in phaeohyceae. This is possible because of the absorption spectra of the phycobilins: the *in vivo* absorption maxima of phycocyanins is at about 620 nm, of phycoerythrocyanins around 545 nm and of allophycocyanins about 650 nm.

Several methods for the purification of this pigment from *Spirulina* sp. grown on a large scale have been described. One procedure for the extraction of phycocyanobilins from *Spirulina* is given here. Cells (wet or lyophilized) are suspended in 10 mM phosphate buffer, frozen and thawed twice and centrifuged (10 000 g for 1 h). The pellet can be used for the extraction of chlorophyll and carotenoids; the blue supernatant, containing the phycocyanin, has to be fractionated stepwise with ammonium sulfate. The precipitate from the 20% saturated ammonium sulfate solution can be discarded, whereas the precipitates from the following 45%, 65% and 75% saturated solutions contain most of the pigment. If a highly purified product is desired, the three fractions can be dialyzed against phosphate buffer and chromatographed on a DEAE-cellulose column with increasing concentrations of phosphate buffer. The yield is about 1% of the dry matter. The quantitative determination of the water-soluble pigments is described in Chapter 9, which deals with estimation of cell growth.

In several cyanobacteria, cyanophycin granules can be found, corresponding morphologically to structured granules, which serve as nitrogen storage compounds in the cells. Cyanophycin consists of multi-L-arginyl-poly-L-aspartic acid, which is also termed as Arg-poly(Asp) or cyanophycin granule polypeptide (CGP), a non-ribosomally synthesized peptide. The molecular weight of cyanophycin ranges from 25 000 to 100 000. It differs from other storage materials as its synthesis occurs largely during the logarithmic growth phase of the algal cells and decreases during the stationary phase, whereas other storage products are usually accumulated at the end of the growth phase or the final stage of the life cycle.

Cyanophycin can accumulate in considerable amounts in stationary-phase cells. For instance, in *Spirulina* sp. about 20% of the protein consists of cyanophycin granules. If the alga is transferred into a nitrogen-free medium, the algal cells continue to grow at their initial growth rate and the amount of total protein does not change, but the concentration of C-phycocyanin decreases at a rate of 30–50%, maintaining the amount of non-phycocyanin protein at a more or less constant level for one cell generation.

The potential for commercial utilization of phycobiliproteins seems to be as natural dyes in the food, drug and cosmetic industries as replacements for synthetic pigments, as highly sensitive fluorescent reagents in diagnostic tests and for labelling antibodies used in multicolor immunofluorescence or fluorescence-activated cell-sorter analysis.

13   Nutrition

13.1.   General introduction

The utilization of different types of SCP including microalgae has a certain tradition in some parts of the world; it was only during the past three decades that the growing market for this new type of product prompted international organizations such as the Protein-Calorie-Advisory Group of the United Nations (PAG) or the International Union of Pure and Applied Chemistry (IUPAC) to publish recommendations and guidelines for the utilization of these unconventional protein sources. In order to give an impression of the quantity and diversity of the recommended evaluations, the major criteria that have to be fulfilled prior to the commercialization of SCP products in general are listed here. These guidelines are intended to serve as recommendations rather than mandatory procedures. In particular, whether the extent of tests with animals and humans is absolutely necessary will depend on the further utilization of the products. The list of evaluations recommended for testing novel sources of SCP is as follows.

1. General characteristics.
   a. Description of the strain of microorganism and its biological properties, assurance of the harmlessness of the organism and the purity of the strain culture.
   b. Characteristics of the substrate and source of nutrient supply and of other agents used in the production process.
   c. Conditions for harvesting and processing.
   d. Constancy and sanitary quality of the product.
2. Product characteristics.
   a. Microscopic morphology.
   b. Physical properties.
   c. Detailed chemical composition.
3. Nutritional studies in test animals (rodents).
   a. Biological value (BV), net protein utilization (NPU).
   b. Protein efficiency ratio (PER), digestibility coefficient (DC).
   c. Digestible and metabolizable energy.
   d. Supplementation properties.
4. Feeding tests in target animal species.
   a. Acceptability.
   b. Evaluation of maximum level of supplementation in normal diets.

196

    c. Tests of possible side effects.
5. Safety evaluations.
    a. Analyses of contaminations with pollutants.
    b. Studies on bacteriological and mycological purity.
    c. Short-term feeding studies with rodents, pigs, birds, etc.
    d. Long-term studies including tests on teratogenicity, carcinogenicity and mutagenicity with two different animal species.
    e. Reproduction studies.
    f. Multigeneration studies.
6. Clinical studies with humans.

The data given in previous chapters on the chemical composition of several algal species have already indicated promising nutritional quality of algal biomass. However, these data alone cannot be considered sufficient to describe the nutritive value. For that, more detailed investigations are required, which determine the availability and digestibility of the nutritive constituents of the algae to the consumer. To study these parameters, a series of animal feeding experiments has been performed by various investigators. The first systematic experiments for the nutritional evaluation of microalgal protein were carried out with rats and chicken in the 1950s and 1960s, but led to quite contradictory results. The obvious reason for the divergent observations were the different methods that have been applied by the researchers to process the algae after harvesting. As will be detailed later, the nutritive value of algae – besides their basic chemical composition – depends on the type of post-harvest drying method used. With the exception of the cyanobacterium *Spirulina*, most of the other types of algae have a relatively thick cellulosic cell wall, which makes the untreated algae indigestible to non-ruminants.

Protein efficiency ratio (PER)

The simplest method, and that most commonly used to evaluate proteins by animal feeding tests, is the determination of the protein efficiency ratio (PER). It is based on short-term (three to four weeks) feeding trials with weanling rats. The response to the diets fed is expressed in terms of weight gain per unit of protein ($N \times 6.25$) consumed by the animal:

$$\text{PER} = \text{weight gain (g)/protein intake (g)}.$$

This method requires only an accurate measure of dietary protein intake and weight gain but needs strict adherence to certain conditions: the calorie intake must be adequate, and the protein must be fed at an adequate but not excessive amount because at high levels of dietary protein, weight gain does not increase proportionately with protein intake. In order to obtain reliable data it is absolutely necessary that the material tested is fully digestible. The greatest source of error in this method lies in the use of weight gain *per se* as sole criterion of protein value. Weight gain cannot be assumed to represent proportional gain in body protein under all conditions.

The PER value obtained is normally compared with a reference protein such as casein. Because of the differences of response to the same standard protein even in the same animal house, the PER values for casein are customarily adjusted to an assumed figure of 2.50, which requires a corresponding correction of the experimental values.

To overcome the shortcomings of the PER estimation, the following more specific methods have been used to evaluate the nutritive quality of algal protein.

Biological value (BV)

Biological value is a measure of nitrogen retained for growth or maintenance and is expressed as nitrogen retained divided by nitrogen absorbed. The absorbed nitrogen is defined as the difference between the ingested and the intestinally excreted nitrogen and can be determined by using the following equation:

$$BV = [I - (F - F_0) - (U - U_0)]/[I - (F - F_0)],$$

where I = nitrogen intake, U = urinary nitrogen, F = fecal nitrogen and F_0 and U_0 are fecal and urinary nitrogen excreted when the animals are maintained on a nitrogen-free or nearly nitrogen-free diet. BV calculated from this equation accounts for metabolic (or endogenous) nitrogen losses. If this correction is not made, that is, if F_0 and U_0 are not considered, the BV obtained is designated 'apparent biological value'. This method is not uncommon and many values for BV represent the apparent biological value of protein.

Digestibility coefficient (DC)

The digestibility coefficient, sometimes called 'true digestibility', expresses the digestibility of the tested protein, i.e. the proportion of food nitrogen that is absorbed by the animal and can be calculated by the following equation, using the experimental data obtained by estimating BV:

$$DC = [I - (F - F_0)]/I.$$

Net protein utilization (NPU)

This parameter can be represented by the simple expression NPU = N retained/N intake. It is equivalent to BV × DC and is a measure both of the digestibility of the diet protein and the biological value of the amino acids absorbed from the food. NPU represents the proportion of food nitrogen retained, whereas BV represents the proportion of absorbed nitrogen required. NPU can be estimated by using the following equation:

$$\text{NPU} = \text{BV} \times \text{DC} = (B - B_k)/I,$$

where B = body nitrogen, measured at the end of the test period on animals fed the test diet, and B_k = body nitrogen, measured at the end of the test period on another group of animals fed a protein-free or low-protein-diet.

Amino-acid nutrition

Animal species differ quantitatively and, in part, also qualitatively in their amino-acid requirement. Data of amino-acid patterns of algal profiles, like those listed in Table 12.2, fail to differentiate between the total amount and the degree of nutritional availability. This aspect, however, is of major importance, especially for the amino acids methionine and lysine. It has been shown that during prolonged storage or thermal treatment of algal biomass, the free amino group of lysine tends to form compounds with reducing carbohydrates (Maillard reaction), making this amino acid non-available for digestion. This effect has to be considered especially with regard to the various drying steps employed for processing the algal material.

Analytical data on the amino-acid composition of a protein allow certain conclusions to be drawn about its nutritional value. However, such considerations presuppose detailed information on the amount of essential amino acids required by the consumer, which differ according to species, age, sex, etc.

Two methods are common for estimating the quality of a given protein by its amino-acid composition: the 'chemical score' and the 'essential amino-acid index (EAA)'.

Chemical score

The chemical score is a simple non-biological device that involves the comparison of the amino-acid composition of the protein to be tested with that of a high-quality protein such as egg, milk or a reference pattern. The score is calculated from the following equation:

$$\text{Chemical score} = \frac{\text{mg of amino acid in 1 g of test protein}}{\text{mg of amino acid in 1 g of reference}} \times 100.$$

The lowest score for any of the essential amino acids designates the 'limiting amino acid' and gives a rough estimate of the quality of the protein tested. In practice, it is preferable for the score for lysine, methionine + cystine, and tryptophane used to be calculated because one of these amino acids is usually limiting in most protein sources.

Essential amino-acid index

The second method, the essential amino-acid index, is based on the assumption that the biological value of a protein is a function of the levels of all the essential amino acids in relation to their content in a reference protein; or in other words, the EAA is defined as the mean of the essential amino acids in the test protein in relation to a complete reference protein (egg).

The test considers the following ten essential amino acids: lysine, tryptophan, valine, isoleucine, leucine, threonine, phenylalanine, methionine and cystine (as one), arginine, histidine.

$$EAA = \sqrt[10]{\left(\frac{lys_{test}}{lys_{ref}} \times \frac{try_{test}}{try_{ref}} \times \frac{val_{test}}{val_{ref}} \times \ldots \times \frac{his_{test}}{his_{ref}}\right)}$$

In general, the chemical score tends to underestimate the protein quality because this method is based on a single limiting amino acid, whereas on the other hand the EAA index results in figures closer to the value determined in biological (feeding) tests. As already mentioned, algal protein is often deficient in methionine and cystine. These deficiencies can be compensated for by supplementing the algal protein either directly with the limiting amino acid or with proteins from other sources rich in these amino acids.

13.2.   Protein efficiency ratio (PER)

Besides a general evaluation of the nutritive quality of algae, this method was especially employed by several authors to demonstrate the influence of post-harvesting treatments on the digestibility of several algal species. To provide an overall view of the major findings, comparative PER values are summarized in Table 13.1. The data represent only a part of numerous reports and give a general idea about the protein efficiency ratio of various algae.

As can be seen from the data given, drum-dried *Scenedesmus* yielded significantly higher PER values than the samples dried by the other methods mentioned. PER was lower at the 20% protein level than at 10%, indicating that 20% already represents a supraoptimal level. In contrast, the sun-dried algae, both cooked and uncooked, showed higher values at the 20% protein level than at 10%, which shows that these methods do not fully make the algal protein accessible for digestion.

When marginal deficiencies in sulfur amino acids were made up by supplementation with methionine, drum-dried *Scenedesmus* gave PER values close to that of casein. The effect of methionine supplementation seems to vary with different algae. It has repeatedly been reported that this supplementation resulted in increased PER for *Scenedesmus* but failed to significantly improve the value of *Spirulina*. Thus, it seems that in the case

Table 13.1. *Comparative protein efficiency ratio (PER) values of different algae processed by different methods*

Alga	Protein level (%)	Processing	PER	Reference
Caesin	10		2.50	Becker *et al.* (1976)
Scenedesmus obliquus	10	DD	1.99	Becker *et al.* (1976)
S. obliquus	10	SD	1.14	Becker *et al.* (1976)
S. obliquus	10	Cooked–SD	1.20	Becker *et al.* (1976)
S. obliquus	10	FD	1.12	Becker *et al.* (1976)
S. obliquus	20	DD	1.68	Becker *et al.* (1976)
S. obliquus	20	SD	1.41	Becker *et al.* (1976)
S. obliquus	20	Cooked–SD	1.52	Becker *et al.* (1976)
S. obliquus	15	SD	0.87	Erchull & Isenberg (1968)
S. obliquus	10	Cooked 8 min	1.78	Pabst (1974)
Chlorella sp.	10	Raw	0.84	Cheeke *et al.* (1977)
Chlorella sp.	10	Autoclaved	1.31	Cheeke *et al.* (1977)
Chlorella sp.	10	DD	1.89	Thananunkul *et al.* (1977)
Chlorella sp.	10	FD	1.66	Lubitz (1962)
Chlorella sp. +0.2% met	10	FD	2.20	Lubitz (1962)
Chlorella sp.	15	SD	0.68	Erchul & Isenberg (1968)
Chlorella sp.	20	Air-dried	1.52	Yamaguchi *et al.* (1973)
Coelastrum proboscideum	10	DD	1.84	Pabst (1974)
Uronema sp.	10	DD	1.43	Pabst (1974)
Oocystis sp.	10	DD	1.39	Mokaday *et al.* (1979)
Spongiococcum sp.	15.3	?	0.94	Leveille *et al.* (1962)
Micractinium sp.	10	DD	2.00	Mokaday *et al.* (1978)
Spirulina sp.	10	SD	1.78	Becker & Venkataraman (1984)
Spirulina sp.	20.5	SD	2.10	Contreras *et al.* (1979)
Dunaliella bardawil	10	DD	0.77	Mokaday & Cogan (1988)

Note:
DD, drum-dried; SD, sun-dried; FD, freeze-dried.

of the latter alga, methionine is not a limiting amino acid in the protein. It is evident that sun-dried *Spirulina* contains an easily digestible protein of good quality because the results obtained were higher than those found for sun-dried *Scenedesmus*, but lower than those of drum-dried *Scenedesmus*. A comparative study between freeze-dried and drum-dried *Spirulina* revealed that the drum-dried material was superior. Nutritional studies with a commercial *Spirulina* product from Mexico and a *Spirulina* sample from Lake Chad, grown in France, showed some differences (Bourges *et al.*, 1971). Whereas a PER of 2.20 was found for the Mexican sample, a value of only 1.86 was obtained for the second one. In earlier investigations performed by Cook (1962), chlorophyceae processed by sun-drying, cooking and freeze-drying gave much lower PER values compared with drum-dried material, no doubt because of the incomplete breakage of the cell wall. Cooking would be expected to improve the PER values; this treatment, however, is rather delicate; if the cooking time is too short it does not rupture the cell wall effectively and if it is extended too ong, it impairs the quality of the algal material.

Summing up the information available, it can be stated that particularly those algae that are generally considered as the most valuable sources of protein (*Scenedesmus, Chorella, Spirulina*) are of good nutritional quality, provided the material is processed by proper treatments and is fully digestible. The protein quality is high compared with other plant proteins and is about 80% of casein.

13.3.   Metabolic studies

The results of the PER studies with different algae have demonstrated the importance of the drying step on the nutritive value. These findings are supported by the results of the various metabolic studies, which are summarized in Table 13.2. In these investigations too, drum-dried *Scenedesmus* sp. was found to have a nutritional quality that is about 85% that of casein; however, the NPU of the alga is lower than for this reference protein, indicating that the unfortified algal protein is limited by at least one of the essential amino acids (methionine). At 10% protein level, all parameters of the drum-dried material were superior compared with all samples dried by other methods. The improvement in the values of the algal protein with a methionine supplement is clearly shown and results from increased utilization of absorbed N and not from higher digestibility of the protein.

The data of the nitrogen balance studies for *Spirulina* confirm that this alga, with its thin and fragile cell wall, does not present serious problems in protein utilization, and even simple sun-drying is sufficient to obtain acceptable values.

Although the various figures differ depending on the algal strain or species tested, it is evident that the algal biomass shows promising qualities as a novel source of protein. Neglecting extreme values, it can be stated that after suitable processing the average quality of most of the algae examined is equal, or even superior, to other conventional high-quality plant proteins. This has repeatedly and unequivocally been confirmed by the long series of different and independent investigations, which analyzed the various metabolic parameters in different animal species.

Table 13.2. *Comparative data on biological value (BV), digestibility coefficient (DC) and net protein utilization (NPU) of different algae*

Alga	Processing	BV	DC	NPU	Reference
Casein		87.8	95.1	83.4	Becker *et al.* (1976)
Egg		94.7	94.2	89.1	Yamaguchi *et al.* (1973)
Scenedesmus sp.	AD	60.6	51.0	31.0	Erchul & Isenberg (1968)
Scenedesmus obliquus	DD	81.3	82.8	67.3	Pabst (1974)
S. obliquus	SD	72.1	72.5	52.0	Becker *et al.* (1976)
S. obliquus	Cooked–SD	71.9	77.1	55.5	Becker *et al.* (1976)
Chlorella sp.	AD	52.9	—	31.3	Bock & Wünsche (1968/9)
Chlorella sp.	DD	71.6	79.9	57.1	Thananunkul *et al.* (1977)
Chlorella sp.	Protein extract	79.9	83.4	66.2	Yamaguchi *et al.* (1973)
Coelastrum probiscideum	DD	75.3	77.8	58.6	Pabst (1974)
Uronema sp.	DD	54.9	81.8	44.9	Pabst (1974)
Spirulina sp.	Raw	63.0	76.0	48.0	Clement *et al.* (1967b)
Spirulina sp.	Stewed	51.0	74.0	38.0	Clement *et al.* (1967b)
Spirulina sp.	SD	77.6	83.9	65.0	Becker & Venkataraman (1984)
Spirulina sp.	DD	68.0	75.5	52.7	Narasimha *et al.* (1982)

Note:
AD, air-dried; DD, drum-dried; SD, sun-dried.

13.4. Protein regeneration studies

A further possibility for evaluating the nutritional quality of proteins is regeneration studies on protein-repleted animals. This procedure is a reliable method for determining the nutritive value of proteins by first depleting and then repleting the protein reserves of adult animals. Significant correlations exist between gain in weight during repletion with the regeneration of blood, liver or carcass proteins, making weight recovery alone a good measure of nutritive value.

The depletion can be accomplished by feeding a protein-free diet to the experimental animals (usually rats) until they have lost about 25% of their initial body weight. For estimations of the nutritive value, seven days of repletion, i.e. feeding of nitrogen in the test protein, are usually sufficient.

This principle of depletion/repletion has been used as an index for evaluating the quality of conventional proteins, but there are very few reports on algal-protein evaluations based on this method. The results of such a study are illustrated by the findings of Anusuya Devi *et al.* (1979, 1983b) for *Scenedesmus obliquus* and *Spirulina platensis*. In these investigations, the animals were divided into five groups. One group was fed on stock diet and served as control without depletion and repletion; the other four groups were fed on a protein-free diet for 12 days, during which the animals lost about 20–25% of their weight. After this period the animals of one group were examined; the remaining rats were fed either on casein (10%) or algal protein (10% and 15%) for a period of 15 days. The intake of algal diets was found to be slightly higher compared with the other diets, whereas weight gain and final body weight of the rats fed casein were higher than in rats fed the algal diets. The major observations are compiled in Table 13.3. It can be seen that the regeneration of liver protein by the casein protein was comparable to that of the algal diet containing 10% protein. The algal protein, however, was less effective than casein for the regeneration of serum albumin. The extent of regeneration of liver enzymes and serum proteins following depletion may be attributed to the quality and level of dietary proteins. Liver enzymes have been shown to be sensitive to protein levels in the diet. Low activity of succinic-dehydrogenase on a protein-restricted diet may be caused by decreased availability of –SH groups for activation, or by a decrease in the absolute amount of the enzyme.

Increased activity of alkaline phosphatase during protein depletion has been reported by others, and attributed to the accumulation of plasma phosphatase in the liver, owing to impaired elimination.

Among the different groups of repleted animals, regeneration of enzyme activity was more pronounced in the casein diet as compared with the algal diets, although the group fed methionine-fortified *Spirulina* reached nearly the same level as the casein group.

13.5.   Digestibility studies

It has already been mentioned that algal proteins (except those originating from cyanobacteria) are poorly utilized when the intact cells are fed to monogastric animals or humans. In order to increase the nutritive availability of these proteins, mechanical, enzymatic, or chemical methods of degrading the algal cells were tested by several investigators. In these studies the digestibility of protein was examined by simulating the intestinal enzyme system pepsin/pancreatin/trypsin in *in vitro* experiments. Selected data from the literature on the effect of the various methods tested are summarized in Table 13.4.

It can be seen that the processing had significant effects on the digestibility of the chlorophyceae; the results of the tests performed by Hindak & Pribil (1968) show a common feature of all the filamentous algae studied in that their proteins were readily digestible *in vitro*, using either pepsin or trypsin or a combination of the two enzymes. It was suggested that the lammelar structure of the cell wall of the filamentous algae probably makes it possible for digestive enzymes to enter the cell interior. In the case of *Spirulina* only marginal differences could be observed between the different treatments tried.

Table 13.3. *Data of protein regeneration studies with rats fed diets containing casein, drum-dried Scenedesmus obliquus and sun-dried Spirulina platensis*

Diet	Weight gained (+) lost (−)	(g per 100 g body weight)			Liver protein (g)	Serum protein (g per 100 ml)			SDH	AAT	AP
		Liver	Kidney	Heart		Total	Albumin	Globulin			
Scenedesmus											
Stock	+85	2.87	0.63	0.29	22.20	7.4	4.8	2.6	1.29	97.84	28.34
Depletion	−36	2.38	0.61	0.36	16.45	4.7	2.8	1.9	0.49	57.44	48.41
Repletion											
Casein (10%)	+83	2.79	0.57	0.33	21.25	7.2	4.7	2.5	0.78	83.07	26.22
Algae (10%)	+45	2.79	0.66	0.35	18.60	6.2	3.4	2.8	0.69	78.52	24.75
Algae (15%)	+60	2.72	0.62	0.31	21.10	6.5	3.6	2.9	0.73	82.75	25.37
Spirulina											
Stock	+84	2.84	0.65	0.30	21.10	7.4	4.8	2.6	1.23	84.23	28.20
Depletion	−34	2.36	0.60	0.38	16.13	4.7	2.7	2.0	0.47	51.62	50.60
Repletion											
Casein (10%)	+82	2.78	0.64	0.34	21.16	7.1	4.6	2.5	0.78	82.42	26.98
Algae (10%)	+56	2.77	0.63	0.22	18.45	6.3	3.4	2.9	0.71	77.75	23.44
Algae (10%) + 0.3% met	+58	2.79	0.65	0.33	18.90	6.6	3.6	3.0	0.74	79.43	24.62

Notes:

SDH, Succinic dehydrogenase (μg TTC reduced per mg fresh liver per 10 min).

AAT, Alanine amino transferase (μM pyruvate liberated per g fresh liver per 10 min).

AP, Alkaline phosphatase (μM phosphorus liberated per g fresh liver per h).

Table 13.4. *Effect of processing on the in vitro digestibility of different algae*

Alga	Treatment	Digestibility (%)		Reference
		Trypsin	Pepsin/Pancreatin	
Scenedesmus obliquus	Fresh	—	30.0	Becker & Venkataraman (1982)
S. obliquus	DD	—	75.0	Becker & Verkataraman (1982)
S. obliquus	Cooked–SD	—	50.0	Becker & Verkataraman (1982)
S. obliquus	SD	—	43.0	Becker & Verkataraman (1982)
S. obliquus	Spray-dried + 2.5% Meicelase	—	54.0	Hedenskog et al. (1969)
Scenedesmus quadricauda	Fresh	—	31.1	Hedenskog et al. (1969)
S. quadricauda	Spray-dried	—	32.2	Hedenskog et al. (1969)
S. quadricauda	Boiled (5 min)	—	33.8	Hedenskog et al. (1969)
S. quadricauda	Fresh	54.0	62.0	Hindak & Pribil (1968)
Chlorella sp.	Fresh	46.2	—	Mitsuda (1962)
Chlorella sp.	FD	65.5	—	Mitsuda (1962)
Chlorella sp.	Protein-isolate	92.7	—	Mitsuda (1962)
Hormidium sp.	Fresh	60.0	84.0	Hindak & Pribil (1968)
Ulothrix sp.	Fresh	88.0	81.0	Hindak & Pribil (1968)
Uronema gigas	Fresh	66.0	88.0	Hindak & Pribil (1968)
Uronema sp.	Fresh	85.0	92.0	Hindak & Pribil (1968)
Stigeoclonium sp.	Fresh	63.0	86.0	Hindak & Pribil (1968)
Spirulina sp.	Fresh	—	82.0	Becker & Venkataraman (1984)
Spirulina sp.	FD	—	70.0	Becker & Verkataraman (1984)
Spirulina sp.	SD	—	65.0	Becker & Verkataraman (1984)
Spirulina sp.	Protein-isolate oven-dried	—	45.0	Al'bitskaya et al. (1979)
Spirulina sp.	Protein-isolate acetone-washed	—	70.0	Al'bitskaya et al. (1979)

Note:
FD, freeze-dried; SD, sun-dried; DD, drum-dried.

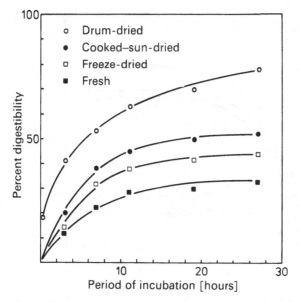

Fig. 13.1. *In vitro* digestibility of differently processed *Scenedesmus obliquus* protein by pepsin and pancreatin.

Protein

Very limited data of *in vitro* studies on the digestion of differently processed *Scenedesmus* and *Spirulina* by simulating the intestinal enzyme system pepsin–pancreatin have been found in the literature. Becker & Venkataraman (1982) reported initial digestibility trials. In their studies, digestibility was determined by first incubating the algae with pepsin for 3 h, followed by pancreatin treatment for 21 h.

Fig. 13.1 shows the protein digestibility pattern of *Scenedesmus*. Hydrolysis reached its maximum level after about 8 h for fresh, freeze-dried, cooked and sun-dried samples. In the drum-dried sample, however, digestion continued up to 24 h and was almost 80%, whereas the fresh alga was least digestible, stressing the important influence of processing and the digestibility of microalgae. For similar *in vitro* studies with *Spirulina*, fresh, freeze-dried and sun-dried material was used. The protein hydrolysis reached its maximum level after about 11 h (Fig. 13.2). It is somewhat surprising that the fresh sample had the highest digestibility, of about 85%, whereas the other two methods of processing reached only 70%. Various other treatments reported in the literature such as boiling with $Ca(OH)_2$, extended autoclaving, sonification, etc. did not significantly alter the digestibility, indicating once more that *Spirulina* biomass *per se* has a high digestibility and does not require further processing.

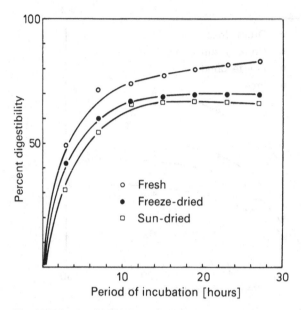

Fig. 13.2. *In vitro* digestibility of differently processed *Spirulina* sp. protein by pepsin and pancreatin.

Carbohydrates

Because it is too costly to extract and isolate algal proteins for utilization, the whole algal cells have to be incorporated in foods and feeds. This means that besides the protein other components of the algal biomass such as carbohydrates, fibers, etc. will affect the overall digestibility. Carbohydrates of algae exist as starch, cellulose, sugars and other polysaccharides; it has been reported that the properties of starch from algae resemble cereal starch or starch in tubers.

In addition to their investigations on protein digestion, Becker & Venkataraman (1982) reported on the *in vitro* digestibility of carbohydrates of drum-dried *Scenedesmus obliquus* and sun-dried *Spirulina* sp. The tests were based on enzymatic amylolysis by α-amylase and subsequent colorimetric estimation of the amount of maltose released by this process.

For *Scenedesmus*, the lowest values were obtained for fresh samples without any further treatment. Cooking increased the amylolysis considerably, whereas autoclaving plus disintegration by glass beads and a hydraulic press prior to amylolysis enhanced the breakdown of starch twofold compared with the uncooked samples (Fig. 13.3).

The extent to which drum-dried *Scenedesmus* samples were hydrolysed by α-amylase is shown in Fig. 13.4. The maximum amylolysis was reached after 3 h of incubation. Though the cooked and autoclaved drum-dried samples gave higher digestibility compared with the uncooked sample, the increase was not significant.

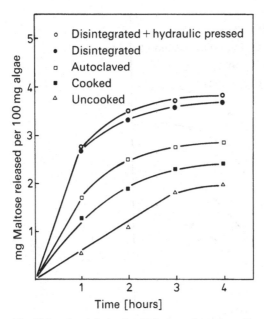

Fig. 13.3. α-Amylolysis of carbohydrate from *Scenedesmus obliquus* after various treatments.

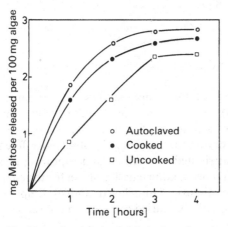

Fig. 13.4. α-Amylolysis of carbohydrate from drum-dried *Scenedesmus obliquus* after various additional treatments.

For 'gastric simulation' prior to amylolysis, HCl and HCl–pepsin treatments were used to preincubate the samples. These treatments effectively release the starch that may be bound and otherwise cannot be acted on by α-amylase. The amylolysis was higher in HCl–pepsin treatments than with the HCl treatment alone in both uncooked and cooked samples. Even for uncooked samples, after HCl–pepsin treatment, the digestibility was higher compared with cooked samples without any pretreatment. Cooking

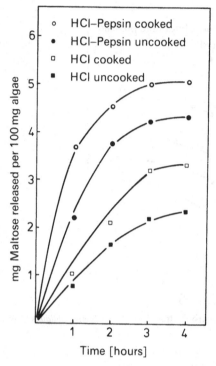

Fig. 13.5. Effect of HCl and HCl–pepsin treatments on α-amylolysis of carbohydrates from *Scenedesmus obliquus*.

improved the carbohydrate digestibility to a great extent in the disintegrated algae (Fig. 13.5).

These results are confirmed by similar studies on the same algae conducted by Soeder (1979), who tested a series of quite different methods of processing, including hot-alcohol treatment, combinations of disintegration, enzyme treatment, drying and cooking. The observed differences in the digestibility could be attributed to swelling and rupturing of starch granules, which facilitates a more random configuration for the enzyme to effect hydrolysis. Higher digestibility of autoclaved samples compared with conventional cooking may partly be caused by rupturing of cell walls and better gelatinization of starch.

The improved digestibility caused by HCl pretreatment prior to amylolysis may be explained by the decrease in the percentage of higher dextrins and malto-triose, resulting in a higher level of reducing sugars. As the digestibility of starch in the carbohydrates is good, there seems to be no limitation in using whole dried algae. Whether carbohydrates are likely to cause any other problems like gastro-intestinal disturbance, flatulence or fluid retention etc., can be established by experiments in the *in vivo* system only.

The studies on α-amylolysis of sun-dried *Spirulina* revealed that there was no detectable maltose release in this alga, even after longer times of incubation. The possible reason may be the low starch content (1%) in the algal material and also the possibility of α-amylase being ineffective against the kind of carbohydrate present in this alga.

13.6.   Supplementation studies

One of the main objectives of producing microalgal biomass was the promise of growing these organisms as an alternative source of protein in feed preparations for various animals species and food formulations for hungry or malnourished populations. In the context of food, it is understood that algae should not serve as the sole source of protein but as a high-protein supplement for diets prepared from conventional agricultural crops.

Several studies have been reported that aimed at improving the nutritive quality of cereal-based protein-deficient diets through supplementation either directly with limiting amino acids, with common proteinaceous ingredients (legumes, skim milk) or with unconventional protein sources such as leaf-protein or single cell proteins.

Over thirty years ago, Hundley & Ing (1956) were probably among the first who tested the feasibility of supplementing conventional food items with algal preparations and amino acids. They reported that the addition of *Scenedesmus* sp. significantly improved the nutritional value of flour and bread, an effect that was attributed to the supplementation of lysine, contributed by the algae. More recently, several authors have studied the properties of supplementing different cereals with varying amounts of *Scenedesmus* sp. and *Spirulina* sp. by feeding these mixtures to rats and estimating PER and NPU values. Representative data of these studies are compiled in Table 13.5. In general, for *Spirulina*, PER and NPU are nearly parallel when they are examined together; the mixture *Spirulina*/corn/oats, however, has a relatively high PER but a relatively low NPU value and the mixture *Spirulina*/wheat (1:3) has a low PER with a comparatively high NPU value. Both mixtures of *Spirulina* with corn had a protein quality higher than corn alone, the effect being more apparent in the mixtures with higher algal content. At a protein concentration of 10%, the utilization seems to be more efficient than at higher amounts, probably because of the fact that the essential amino-acid pattern of the 10% protein diet is already close to the pattern of a well-balanced protein.

Rice, wheat, ragi (*Eleusine coracana*) and other millets form the major protein sources of the diets of rural communities in several poorly nourished populations. Experiments have demonstrated that it is possible to improve the quality of these cereals that are often deficient in lysine and threonine by supplementation with algae. To quantify the supplementing

Table 13.5. Data on the supplementary value of different algae

Diet	Protein	PER (%)	NPU	Reference
Casein	10.0	2.50	54.4	Bourges et al. (1971)
Corn	7.0	1.23	30.5	Bourges et al. (1971)
Wheat	9.0	1.43	32.5	Bourges et al. (1971)
Spirulina + corn (1:1)	10.0	1.72	34.7	Bourges et al. (1971)
Spirulina + corn (3:1)	10.1	1.80	37.2	Bourges et al. (1971)
Spirulina + wheat (1:1)	14.5	1.96	48.4	Bourges et al. (1971)
Spirulina + wheat (1:3)	10.8	1.74	41.1	Bourges et al. (1971)
Spirulina + corn + oat (5:3:2)	10.0	1.91	35.0	Bourges et al. (1971)
Spirulina + corn + rice (2:2:1)	10.0	1.95	45.1	Bourges et al. (1971)
Spirulina + rice (1:1)	9.0	2.27	—	Anusuya Devi & Venkataraman (1983a)
Spirulina + rice (1:3)	8.0	2.21	—	Anusuya Devi & Venkataraman (1983a)
Spirulina + wheat (1:1)	10.0	1.74	—	Anusuya Devi & Venkataraman (1983a)
Spirulina + wheat (1:3)	10.0	1.34	—	Anusuya Devi & Venkataraman (1983a)
Barley	10.0	—	58.0	Narasimha et al. (1982)
Spirulina	10.0	—	52.7	Narasimha et al. (1982)
Spirulina + barley	10.0	—	61.2	Narasimha et al. (1982)
Rice	5.8	2.05	—	Venkataraman et al. (1977b)
Ragi (Eleusine coracana)	6.0	1.15	—	Venkataraman et al. (1977b)
Wheat	10.0	1.13	—	Venkataraman et al. (1977b)
Scenedesmus	10.0	1.99	—	Venkataraman et al. (1977b)
Scenedesmus + rice (1:1)	9.3	2.44	—	Venkataraman et al. (1977b)
Scenedesmus + rice (1:3)	8.5	2.21	—	Venkataraman et al. (1977b)
Scenedesmus + ragi (1:1)	8.8	2.04	—	Venkataraman et al. (1977b)
Scenedesmus + ragi (1:3)	8.6	1.75	—	Venkataraman et al. (1977b)
Scenedesmus + wheat (1:1)	10.0	1.90	—	Venkataraman et al. (1977b)
Scenedesmus + wheat (1:3)	10.3	1.52	—	Venkataraman et al. (1977b)
Scenedesmus + wheat + rice (1:1:1)	9.5	2.20	—	Venkataraman et al. (1977b)

effect of drum-dried *Scendesmus* and sun-dried *Spirulina*, several combinations of algae and conventional food ingredients have been tested by carrying out PER experiments. As judged by the rat growth experiments, both algae showed promising supplementary values to the cereals. It is interesting to note that ragi proteins, having a low PER value, are considerably improved nutritionally by the addition of algae. In the case of wheat, marked improvement of the PER was observed when its proteins were blended with algae proteins in the ratio of 3:1. The results also indicate that supplementation with algae improved the PER of ragi- and wheat-based diets to a greater extent than rice; a blend of equal amounts of *Scenedesmus*, rice and wheat gave the highest PER.

There is no doubt that the addition of algae has a positive effect on the nutritional quality of many common proteins. However, the question remains as to the form in which the consumer can make use of the promising attributes of the algae; serious acceptability problems will exclude the distribution of plain algal powder to the consumer. To overcome these obstacles and to popularize algae as a food supplement, the material has to be incorporated into common recipes and food preparations that are typical for the particular region.

One of the numerous possibilities in this connection is the incorporation of algae into bread and other baking products. As an initial step in this direction, preliminary baking and extruding trials have been performed in order to study the possibility of adding algae to enrich the protein content of bread (Nigam *et al.*, 1985). Wheat flour and whole-wheat flour were used for baking trials; the extrusion experiments were conducted with whole-wheat flour alone. Drum-dried *Scenedesmus obliquus* and sun-dried *Spirulina platensis* were added at 5, 10 and 15% levels. In the baking tests, the yield of both dough and bread increased with the amount of algae added although the bread volume decreased slightly when *Scenedesmus*, and considerably when *Spirulina*, was added. The porosity of the bread remained unaffected, but there was a strong discoloration of the crumb in the case of both wheat flour and whole-wheat flour bread, making the product less acceptable for the consumer. In the extrusion trials, comparatively good specific volume and only slightly discoloration were obtained at a moisture content of 14% and an algal content of 5%. At higher moisture (25%) and algal contents (10, 15%), the color of the product became greenish-brown and less appealing.

The use of dried algal powder at levels of 5–15% in traditional foods has been shown to be safe. Despite this, few preparations, such as noodles, ravioli, fruit pudding, protein and vitamin food supplements and health foods containing algae have been developed. It was assumed that the dark color imparted by the algae is due to its physical state like its particle size. In order to better analyze this effect, studies were performed to describe the influence of particle size on the visual color response by examining the individual powdered algal fractions (*Scenedesmus* sp. and *Spirulina* sp.) before and after mixing with wheat flour (Nigam *et al.*, 1990).

Color analysis by means of a colorimeter revealed that reflectance from similar-size fractions of algal powder was higher for *Spirulina* by a value of about four as compared with *Scenedesmus*. It was found that to increase reflectance significantly and thus produce correspondingly less-colored products, the most suitable algal particle size is between 10 and 40 μm. Incorporation of water as in dough preparation is expected to destroy the mirror effects, resulting in a decrease of reflectance and subsequent darkening of the product. Nevertheless, in conclusion, a comparatively lighter-colored baked and extruded wheat product with less final moisture content than a conventional product results with algal particle diameters between 10 and 40 μm. This is especially true if *Spirulina* is incorporated at the 5% level. *Spirulina* gives less color compared with *Scenedesmus* at the same particle size.

It has already been mentioned that the fortification of algal-based diets with certain essential amino acids improves the quality of the algal protein, resulting in PER values close to that of casein. This observation was supported by experiments designed to determine the feed conversion efficiency (FCE = feed consumed (g)/weight gain (g)) of *Chlorella*, fortified by the addition of methionine (0.3%) and lysine (0.3%), and soybean meal as reference protein (Cheeke *et al.*, 1977). Whereas the unfortified alga gave an FCE of 5.5, the addition of lysine resulted in growth (FCE = 3.8) which was equal to that obtained for soybean meal. When both amino acids were added simultaneously, the FCE further decreased to 3.6, which is even better than that of the reference. It appears that, according to chemical amino-acid determination, methionine is the first limiting amino acid; in practical feeding, however, lysine is limiting. Because algae are generally high in lysine content, it is necessary to find out why lysine availability is low. Inactivation during processing (Maillard reaction) and/or lysine 'tie-up' as part of the indigestible cell wall are two possibilities that have to be considered in this case.

Other authors found differing results on the supplementation of *Scenedesmus* sp. with amino acids (Gross *et al.*, 1982) and reported a continuous decrease of the PER with increasing quantities of methionine, starting at 0.1% in the diet (Table 13.6). A possible explanation for this adverse effect might be the fact that feed preparations with higher concentrations of methionine can get a bitter taste, resulting in poorer palatability and lower consumption of the diet. No significant differences were found by the authors on possible interactions of methionine and isoleucine on PER and feed intake.

A detailed evaluation of limiting amino acids and digestibility of *Spirulina* sp. has been reported by Hernandez & Shimada (1978) based on

Table 13.6. *Effect of different amino-acid fortifications on the PER of different algae*

Diet	Protein (%)	PER	Reference
Casein	10.0	2.50	
Scenedesmus	10.0	2.20	
Scenedesmus + met (0.05%)	10.0	2.20	Gross *et al.* (1982)
Scenedesmus + ile (0.05%)	10.0	2.25	
Scenedesmus + ile (0.05%) + met (0.05%)	10.0	2.20	
Spirulina + met (0.29%)	19.6	1.35	
Spirulina + met (0.29%) + lys (0.27%)	19.6	1.21	
Spirulina + met (0.13%) + his (0.13%)	22.0	1.65	Hernandez & Shimada (1978)
Spirulina + met (0.13%) + his (0.13%)	22.0	1.40	
Casein + lys (0.2%) + thr (0.1%) + met (0.25%)	10.0	2.50	
Dunaliella	10.0	0.77	
Dunaliella + lys (0.5%) + met (0.25%)	10.0	2.20	
Dunaliella + met (0.25%) + ileu (0.07%) + try (0.05%)	10.0	1.33	Mokady & Cogan (1988)
Dunaliella + lys (0.5%) + ileu (0.07%) + try (0.05%)	10.0	1.55	
Dunaliella + lys (0.5%) + met (0.25%) + ileu (0.07%)	10.0	2.15	
Dunaliella + lys (0.5%) + met (0.25%) + try (0.05%)	10.0	2.00	

rat feeding experiments. None of the amino acids tested (methionine, histidine and lysine) added alone or in combination to diets containing 10% protein provided exclusively by *Spirulina* seems to be limiting for this alga. The comparatively low PER values might be a result of amino-acid imbalances caused by adding pure amino acids, which are absorbed much more quickly than those bound in proteins.

In recent years, the green alga *Dunaliella* has been the subject of wide attention, mainly because of its unique ability to produce and accumulate high concentrations of β-carotene and glycerol. Following extraction of these two valuable chemicals, a residue containing up to 65% is obtained. The nutritional quality of this protein was evaluated by Mokady & Cogan (1988) using the protein efficiency ratio (PER) method. These studies revealed a very low PER value for *Dunaliella* of only 0.77, indicating that the protein is limited in several essential amino acids. Because amino-acid analyses of the protein showed inferior concentrations of isoleucine, tryptophan and lysine, as compared with the reference soya protein, several animal feeding tests were conducted by supplementing the *Dunaliella* protein with the deficient amino acids. As can be seen from Table 13.6, the addition of these amino acids in various combinations substantially improved the PER to values similar to that of supplemented casein. The remaining differences between supplemented control and supplemented algal protein might have arisen from an imbalance between leucine and isoleucine in *Dunaliella* that could not be overcome by supplementation with isoleucine and that is known to reduce protein quality.

In conclusion, algal proteins have a nutritionally almost balanced amino-acid profile – with marginal deficiencies in sulfur-containing amino acids – and often supraoptimal concentrations of lysine. Combinations of algae with cereals and other basic food ingredients would serve to improve the nutritional quality of several staple foods and at the same time could prevent acceptability and tolerance barriers of algal proteins, thus providing a sound base for the industrial production of algal-based food preparations.

13.7.   Toxicological studies

Introduction

Before a new food item is declared safe for human consumption, it has to undergo a series of detailed toxicological tests to prove the harmlessness of the product. This applies especially to unconventional protein sources, which is where microalgae are grouped. Besides chemical and sanitary analyses, these toxicological evaluations include short- and long-term feeding studies with rodents and other animal species, multi-generation studies, teratogenic and mutagenic studies as well as pre-clinical and clinical studies with humans.

Recommendations for the performance of such evaluations have been published by different international organizations (Protein Advisory Group, 1972). However, it has to be assumed that additional national laws and regulations exist, which specify the recommended tests.

Whether or not a particular algal biomass is suitable for utilization as feed or food is determined by several factors related to the nutritional quality and the toxicological safety. The safety of all kinds of unconventional protein sources will depend on the organisms selected, the quality of the substrate used and the conditions of growth. Analyses of chemical composition of algae and results of nutritional studies have been described previously. This section deals with safety evaluations, sanitary analyses and a few preliminary results of clinical and acceptability studies.

As part of the toxicological characterization of the algal biomass the material has to be analyzed for the presence of toxic compounds, either synthesized by the algal cell or accumulated from the environment. These toxins can be grouped into two categories, i.e. biogenic and non-biogenic toxins.

According to this classification, the first group includes all those compounds that are either synthesized by the cells or formed through decomposition of metabolic products. The second group comprises environmental contaminants and other substances, mostly of anthropogenic origin, that enter the algal culture from outside and are absorbed and accumulated by the algal cells. Whereas the occurrence of biogenic toxins is an intrinsic characteristic of the organism itself, contaminations with non-biogenic toxins can be avoided in most cases by proper cultivation techniques, plant management and selection of cultivation areas that are free of pollution.

Biogenic toxins

Nucleic acids

One of the very few algal constituents that, under certain circumstances, may be counted as biogenic toxins are nucleic acids. Because these compounds are repeatedly accused as the major limitation in the use of algae and other microorganisms as food sources, it is felt that some reflections are necessary at this point to illustrate the matter.

The nucleic-acid content is approximately 4–6% for algae, 8–12% for yeasts and up to 20% for some bacteria. Sources of purines in the human diet consist of ribonucleic acid (RNA) and deoxyribonucleic acid (DNA). The ingestion of purines may lead to an increase of plasma uric-acid concentration and urinary uric excretion due to the inability of the human metabolism to degrade purines beyond the level of uric acid. Elevated serum levels of uric acid increase the risk of gout, while increased urinary concentration of uric acid may result in the formation of uric acid stones in the kidney, nephropathy, or other health hazards.

The normal plasma uric-acid concentration of men is 5.1 ± 0.9 mg 100 ml^{-1}, the plasma uric-acid concentration of women is about 1 mg lower than that of the male. Standards for children are not so well established but appear to fall in the lower range of the values for adults. Most authorities accept 6.0 mg of uric acid per 100 ml plasma as the lower limit of the high-risk population. The daily ingestion of 1 g of RNA per 70 kg body weight will increase the serum acid concentration by 0.9 mg 100 ml^{-1} and the urinary uric acid by 100–150 mg, indicating that only part of the purines in this nucleic acid is absorbed and eventually excreted as uric acid by the kidney. The effect of DNA on uric acid level in both urine and plasma is less than half that of RNA. The ratio of RNA:DNA in algae is about 3:1.

Because of a possible health hazard, the Protein Advisory Group recommended that the daily nucleic acid intake from any unconventional source should not exceed 2.0 g, with total nucleic acid from all sources not exceeding 4 g per day, thus setting a safety limit. It is further stated that the serum uric-acid concentration should not exceed 8 mg 100 ml^{-1}. Existing information on uric acid levels in blood and urine in relation to diets has been derived from studies on populations consuming West European and North American diets.

In order to collect data about the uric-acid level in Asian populations, estimations on the uric acid concentration of the population in Thailand have been conducted by an experimental algal plant at Bangkok. The reported average value for women was 4.3 mg 100 ml^{-1} and for men 5.75 mg 100 ml^{-1}. By considering the standard deviation and other statistical parameters, the standard value for women would be 2.25–7.0 and for men 3.4–7.8 mg 100 ml^{-1}. Data on the effect of algae-containing diets on the uric-acid concentrations in humans are scarce. Experiments conducted with volunteers did not reveal an alarming increase of uric-acid concentration due to the consumption of algae. This was confirmed by Feldheim (1972) who reported the results of a study performed in Thailand where 20 adult persons received a normal diet with the addition of 21 g of drum-dried *Scenedesmus* (7 g per meal) and another 20 persons received the same diet without the addition of algae. It was found that the consumption of the algae did not change the concentration of uric acid normally present in the tested individuals (4.9–5.5 mg 100 ml^{-1}). Higher algal concentrations were used by Waslien *et al.* (1970) who determined uric-acid levels in men fed algae as the sole source of protein. Seven healthy men were given casein- or algae-containing diets at two dietary levels to provide 25 and 50 g of protein daily; pure RNA was added to the casein. In a purine/protein free diet the mean uric acid concentration in the plasma was 5.4 mg 100 ml^{-1}. After a daily intake of 25 and 50 g of protein from casein, providing 1.8 and 3.7 g of RNA, the uric acid concentration increased to 6.9 and 8.7 g 100 ml^{-1}, respectively. The intake of equal amounts of algal protein (1.7 and 3.6 g of nucleic acid) elevated the plasma uric-acid concentration to 7.4 and 9.7 mg 100 ml^{-1}, respectively, thus reaching the limit of the pathological range.

Such elevated plasma uric-acid levels could not be substantiated by Griebsch & Zöllner (1970) who examined the effect of drum-dried *Scenedesmus obliquus* on uric-acid metabolism. The authors administered an algal concentration, which provided the minimum requirement to maintain protein balance, i.e. 0.62 g of protein per kg body weight as well as one half and double this amount (0.31 and 1.24 g per kg body weight). At the lowest intake (0.31 g kg^{-1}) uric-acid levels, even in 'latent' hyperuricemic persons, remained within the norm (6.5 mg 100 ml^{-1}), whereas the level increased to 13 mg 100 ml^{-1} when 1.24 g of protein per kg body weight were consumed daily. Somewhat different is the metabolization of nucleic acids in certain animal species. Pigs have a very low excretion capacity for uric acid (ten times less than man), which, however, is well compensated for by their ability to build high amounts of allantoic acid, which is harmless in respect to health disturbance. Therefore, it can be expected that pigs may tolerate higher amounts of algae in their diets than man. Also, broiler production on algal mass presents no problem from the standpoint of nucleic acids because poultry have a high excretion capacity for uric acid as related to body weight. Even fewer problems can be expected with fish, which transform most compounds of nitrogen to urea and ammonia.

Phaeophorbide

In view of the utilization of algae as a food substitute, some of the findings reported from Japan should be mentioned here.

In 1977, it was observed that humans developed photosensitized skin inflammation after the consumption of *Chlorella*. These irritations were caused by the chlorophyll degradation products phaephorbide and its esters formed by the peroxide formation of fatty acids (arachidonic acid) in the cell membrane. The minimum effective dose was reported to be a daily intake of 25 mg of phaeophytin. These dermatis-like symptoms could also be provoked in rodents in laboratory experiments. Animal studies with rats showed that the LD_{50} for phaephorbide$_a$ was 455 mg kg^{-1} body weight and the MLD 120 mg kg^{-1} body weight.

The first step in the formation of phaeophorbide, which is present in small amounts in all chlorophyll-containing parts of plants, is the removal of the magnesium, the central atom of chlorophyll, initiated by the action of dilute mineral acids, leading to the formation of phaeophytin. Further removal of the phytol ester on the C-7 propionate group of the chlorophyll molecule by the enzyme chlorophyllase results in the formation of phaeophorbide. In this context it has to be observed that chlorophyllase, to a certain extent, withstands thermic treatment and thus remains viable in algal cells dried at a moderate temperature. Furthermore, its activity is enhanced by humidity, resulting from insufficient drying or improper storage. It has also been established that ethanol used during the processing of *Chlorella* enhances the chlorophyllase activity.

To inactivate this enzyme, the Japanese Ministry recommended intensive heat treatment of 100 °C for at least 3 min. It has been reported for *Spirulina* (Jassby, 1988) that there is a distinct seasonal pattern in the pheaophorbide level related to the ambient temperature owing to activation of chlorophyllase activity. The alarming reports on these side effects caused by the consumption of insufficiently dried algae prompted the Japanese Ministry of Health in 1981 to publish instructions that the concentration of total phaeophorbides in *Chlorella* products sold commercially has to be below 1.2 mg g^{-1}.

Phycotoxins

Although this book deals mainly with the positive qualities of algae that place them on the black side of the ledger, there a few species among the microalgae that must be listed in red as liabilities because they are forms that produce different types of phycotoxins. When we propose to culture algae for food or feed we must contend with those algae that produce toxins and may contaminate food cultures.

Man has been poisoned by algal toxins, and poisoning of livestock and other animals attributed to toxic blooms of algae occur with considerable frequency but unpredictably in comparable climatic and geographic regions of South America, Europe, Asia, South Africa and Australia. These poisonings can cause considerable economic loss and they are also of concern to wildlife biologists, environmentalists and public health officers.

Because there has been concern about the possibility that some of the algal species selected for mass cultivation may produce toxic compounds, a brief review on the toxicity of certain algae shall be given here. A collection of the important algae that have been classified as potentially toxic is listed in Table 13.7.

Laboratory experiments including intravenous, intraperitoneal and oral exposure with extracts of different algae have been performed on laboratory animals with death occurring from within a few minutes to as much as 48 h. Therefore the different toxins are classified into three major groups, i.e. VFD = very fast death factor, operating within 3 min; the FDF-fast death factor; killing within 2 h; and the SDF = slow death factor causing death within 4–48 h.

It has been reported repeatedly that in fresh-water lakes, abundant algal blooms may lead to death of fish either directly by algal toxins liberated into the water or indirectly by by-products of algal decay (for instance, the protein derivative hydroxylamine).

The algae responsible for these incidences are mainly species of *Aphanizomenon* and *Microcystis*. Whereas the toxin of *Aphanizomenon* is actually a mixture of different toxins, one of which is saxitoxin, the toxin of *Microcystis* is a polypeptide containing equimolar amounts of methionine, tyrosine, alanine, glutamic acid, erythromethylaspartic acid and methylamine and it behaves as a neurotoxin. The effects produced in animals are said to be much the same as those caused by the toadstool *Amanita phalloides*. Species of marine dinoflagellates such as *Gymnodinium*, *Gonyaulax* and *Pyrodinium* are infamous perpetrators of fish and shellfish deaths. Spectacular deaths of marine fish and mussels occur in the so-called 'red tide', a bloom of the alga *Gymnodinium*. These red tides mainly occur along the coasts of Florida, New Jersey, California and the Gulf of Mexico. It has been described that even air that blows landward from red tide waters, produces respiratory ailments among humans. Death or poisoning of humans resulting from the consumption of fish and shellfish are also caused by the above dinoflagellates (especially *Gonyaulax catenella*), at least indirectly. The toxins are harmless to the fish or shellfish (*Mytilus*) themselves, where they are bound to the dark gland or the hepatopancreas, without causing observable disturbances. As the human is the last member of food chain, these toxins accumulate in the human tissues, especially the liver, and cause severe illness or death. The toxin produced by *Gonyaulax* is estimated to be ten times more potent than strychnine. Several exotoxins and endotoxins, produced by fresh-water cyanobacteria, can bring about the death of farm animals. The most toxic of these compounds are the exotoxins produced by *Microcystis* sp., but *Anabaena* and *Aphanizomenon* are also known to be toxic. Animals such as cattle, horses, sheep and even birds are killed or made seriously ill by drinking water infested with these algae. It has been reported from South Africa that blooms of *Microcystis* sp. in the shallow lakes of the grazing range cause the loss of thousands of cattle annually, manifested by loss of weight, weakness, liver pathology and abortion. An additional hazard is that the pigment phycocyanin from the algae is carried to the skin of the animals where it is sensitized by the light resulting in internal burning and peeling of the skin. Because of the concern that those algae that are used presently for large-scale production may also contain certain toxins, either produced by the algae themselves or by other accompanying microorganisms always present in outdoor cultures, detailed analyses have been performed in Germany for the presence of toxins in samples of *Scendesmus obliquus* and *Spirulina platensis*. The following compounds have been sought: aflatoxin, ochratoxin A, sterigmatocystin, citrinin, patulin, penicillic acid, zearalenone, diacetoxyscirpenol and thrichothecene. In addition, dermatological toxicological tests with trichothecene, toxicity studies with mice on tremorgens (fumitremorgen and verrucolugen), and investigations on antibiotic activities (penicillium toxins) have been conducted. None of the compounds could be detected and all biological tests were negative, indicating that the two algae were free from toxins.

Based on the many nutritional and toxicological studies performed so far with these algae, it can be stated that they are free of any phycotoxins. However, it is always advisable to proceed with caution because there exist toxic strains of cyanobacteria that are morphologically indistinguishable from non-toxic strains and that may contaminate cultures of other harmless algae.

Non-biogenic toxins

Among the non-biogenic compounds, special attention has to be given to the concentration of heavy metals and polycyclic aromatic compounds in the algal biomass.

Table 13.7. *List of selected toxic algal species, their phycotoxins and possible hazards*

Algal species	Toxic property	Toxic compound	Toxicated organism
1. CHLOROPHYTA			
Marine			
Caulerpa racemosa		caulercipin	mice, men
C. sertularoides		caulerpin	
C. lentillifera		(hydroxy amide)	
C. lamourouxii			
Chaetomorpha minima	haemolytic	fatty acids	fish
Ulva pertusa	haemolytic	galactolipids	sea urchin egg
2. CYANOPHYTA			
Marine			
Lyngbya gracilis	dermatitis	phenolic compound	
L. majuscula		debromo-aplysiatoxin	fish, rat, men
Schizothrix calcicola	gastrointestinal		
Oscillatoria nigrovidis	disorder, diarrhhoea tumor promotor	oscillatoxine	
Fresh water			
Microcystis aeruginosa	diarrhoea convulsion	microcystine	mice, men
Anabaena flos-aquae	paralysis	alkaloid anatoxin	poultry
Aphanizomenon flos aquae	paralysis	saxitoxins	cattle, fish
Synechococcus sp.	neurotoxic	pteridinlike	mice, fish
Coelosphaerum sp.			
Nodularia sp.			
Gloeotricha sp.			
Trichodesmium sp.			
Nostoc rivulare	carcinogenic		

3. CHRYSOPHYTA
Fresh water

Prymnesium parvum	haemolytic	glycolipid	fish
Ochromonas malhamensis	antispasmotic	lipoprotein	
O. danica			fish
O. minuta	haemolytic		fish
O. sociabilis			
Apistonema sp.	haemolytic		fish
Pleurochrysis scherffelii			

4. PYRRHOPHYCOPHYTA
Marine

Gonyaulax catenella	paralytic shellfish poisoning and fish poisoning	saxitoxins	fish, men
G. acatenella			
G. tamarensis			
G. polygramma			
G. monilata			
G. polyedra			
Noctiluca miliaris		ammonia	fish
Pyrodinium phoenus			
Diplosalis sp.	nephrotixic		fish
Peridinium polonicum		alkaloid	fish
P. veneficum			
Aphidinium kleosii		cholin esters	
A. rhynchocephalum			fish
A. carteri			

5. PHAEOPHYTA
Marine

Pelvetia fastigiata		fastigiatine	

Heavy metals

It is a well-known fact that practically all microorganisms are capable of accumulating heavy metals at concentrations that are several orders of magnitude higher than those present in the surrounding media. This metal accumulation is a relatively rapid process; bacteria, for instance, can attain an equilibrium distribution of heavy metals between the cell and the surroundings within a few minutes. In the case of algal cells, saturation with heavy metals normally will be reached within 24 h; any exposure time longer than that will not increase the accumulation capacity of the cell. Thus, the consequence for large-scale cultivation is that the contamination of algae with heavy metals, if they are present in the medium, cannot be avoided, because the average cultivation time is longer than the time required for maximum metal uptake. Elevated amounts of various heavy metals in the algal biomass was, and in certain areas still is, one of the major problems that limit the large-scale utilization of algae. According to the guidelines published by WHO/FAO, an adult person of 60 kg body weight should not consume more than 3 mg of lead, 0.5 mg of cadmium, 20 mg of arsenic, and 0.3 mg of mercury per week. Because children react much more sensitively to these heavy metals than adults, the tolerable amounts for children have to be still lower than calculated based on body weight alone.

No typical or characteristic levels exist for toxic metals in microalgae; the concentrations are very variable, even for products from the same site, which may be explained by varying composition of the culture media, contamination during processing, or even improper analytical techniques.

At present, no official standards exist for the heavy-metal content of miocroalgal products. However, on a voluntary basis some algae manufacturers, for instance a Japanese *Chlorella* producer, have established internal guidelines for metal levels in their products (Jassby, 1988).

To illustrate the present situation concerning the levels of heavy metals found in algal samples, selected analytical data found in the literature are summarized in Table 13.8 and compared with recommendations on upper limits of heavy-metal contents in different food commodities.

It can be seen that several of the concentrations detected in the algal samples exceed the recommended limits. The main causes of these high levels are environmental pollutants due to unfavourable locations of the algal plants as well as contaminations introduced through water and fertilizer.

Besides the amount of heavy metals accumulated by the algae during cultivation, there are other possibilities by which these impurities can be introduced into the agal biomass. Under alkaline conditions and in the presence of phosphate and sulfate ions from the fertilizer, dissolved cadmium and lead ions in the medium form slightly soluble compounds that precipitate or float adhering to small particles. These particles will be

Table 13.8. *Data on heavy-metal concentrations in algae compared with international recommendations (data given p.p.m.)*

Source	Pb	Cd	Hg	As	Reference
Maximum weekly intake (mg per adult)	3.0	0.5	0.3	20.0	WHO (1972)
Limits in SCP	5.0	1.0	0.1	2.0	IUPAC (1974)
Limits in drinking water	0.1	0.01	0.001	0.05	WHO (1972)
Japan Chlorella Industry (for *Chlorella*)	2.0	0.1	0.1	0.1	Jassby (1988)
Japan Health Food Ass. (for *Spirulina*)	20.0 as total lead				Jassby (1988)
Scenedesmus (Thailand)	6.03	1.67	0.07	2.36	Payer & Runkel (1978)
Scenedesmus (Germany)	34.8	2.46	0.09	2.36	Payer & Runkel (1978)
Scenedesmus (Peru)	0.58	0.30	0.43	0.91	Becker & Venkataraman (1982)
Spirulina (India)	3.95	0.62	0.07	0.97	Becker & Venkataraman (1982)
Spirulina (Mexico)	5.1	0.5	0.5	2.9	Boudene *et al.* (1975)
Spirulina (Chad)	3.7	—	0.5	1.8	Boudene *et al.* (1975)

harvested together with the algal biomass and subsequently will be found in the algal material after processing. This risk of contamination can be reduced by rinsing the harvested algal slurry with slightly acidic water or with complexing agents in order to dissolve and remove the heavy metal phosphates and sulfates.

The literature dealing with toxicological effects of heavy-metal-containing algal on the consumer is very small. One of the few reports considering this aspect is the study of Yannai *et al.* (1979) who tested the quality of highly contaminated waste-water-grown algae (*Scenedesmus* sp., *Micractinium* sp. and *Chlorella* sp.) by feeding them to carp and poultry. The chicken were grown on diets containing 15%, the carp on diets containing 25% dried algae (Table 13.9).

No significant accumulation of any of the analyzed elements (mercury, copper, cadmium, lead, aluminum and arsenic) could be detected in the tissues of chicken or carp in spite of the high concentration in the diets. This observation may be explained by the unusually large amount of phosphate present in the algal material, because all the heavy metals appeared to form water-insoluble phosphates that could not be absorbed by the gastrointestinal tract of the animals. This assumption was strengthened by another study in which sewage-grown *Micractinium* sp. with elevated levels of heavy metals was fed to Japanese quails. Here too, the percentage of metal accumulation by the animals was very low. In addition, the high amount of aluminum, which was present in the algae as result of the aluminum flocculant used for harvesting, did not affect the performance of the birds.

Table 13.9. *Correlation between heavy-metal concentration in algal diets and levels found in tissues or chicken fed on these diets (p.p.m.)*

	Alga	Liver	Leg Muscle	Tibia
Scanedesmus				
Hg	0.30	0.04	0.04	—
Cu	45.30	3.18	1.12	1.89
Cd	1.6	0.24	0.12	0.60
Pb	3.8	0.45	0.40	3.70
Al	33.9	—	—	1.4
As	1.1	0.81	0.11	—
Micractinium				
Hg	0.64	0.06	0.047	—
Cu	33.10	3.20	0.83	2.81
Cd	1.3	0.16	0.07	1.49
Pb	2.9	0.37	0.27	0.44
Al	7.4	—	—	6.40
As	1.3	1.46	0.18	—
Chlorella				
Hg	0.26	0.07	0.06	—
Cu	24.20	3.20	0.84	2.39
Cd	1.4	0.16	0.08	1.40
Pb	8.1	0.3	0.34	4.70
Al	0.33	—	—	5.10
As	3.6	1.07	0.21	—

As a continuation of this study, a second investigation was performed in which weanling rats were fed on meat from the chickens that were raised on the contaminated algae. After 12 weeks these rats were mated and the offspring were raised on the same diet. At the end of the experimental period, no significant difference in growth, performance, behavior, survival rate, fertility or lactation could be detected between the animals grown on the test diet and those on a control diet. Histological examination of various tissues did not show any abnormalities in the experimental animals. Similar results were obtained by other authors who fed heavy-metal-contaminated *Spirulina* sp. to rats over a period of 75 weeks (Boudene *et al.*, 1975). The test animals showed no difference from the controls; the growth rate was comparable and no evident toxicity could be detected.

To minimize the risk of heavy-metal contamination, high-grade nutrients should be used as fertilizer, because phosphorus fertilizer especially contains unacceptable levels of various heavy metals. Furthermore, certain polyvinylchloride plastic linings used for the construction of alga ponds contain lead as a stabilizer and/or as a pesticide to prevent degradation by soil bacteria.

In conclusion it has to be stressed once more that, especially if the algal biomass is intended to be used for feed and food purposes, care has to be taken to keep the amount of heavy metals in the algae as low as possible. It seems that even today, where this problem has been recognized, many algal charges do not fulfil the recommended quality criteria regarding the permitted heavy-metal contents.

Organic compounds
Another group of non-biogenic hazardous pollutants found in algal samples include organic compounds such as polychlorinated biphenyls and polycyclic aromatic hydrocarbons. All of them are of anthropogenic origin and several are highly toxic or known carcinogens. Very few analytical data on the concentrations of some of these chemicals in algal samples could be found in the literature, mainly because their qualitative and quantitative determination requires expensive analytical methods.

Table 13.10 summarizes some of the published results. It should be stressed that the figures are no more than a crude orientation because the amount of these chemicals accumulated in the algae depends on environmental factors as well as on the degree of pollution in the surroundings of the algal production units. Naturally, these contaminants differ from place to place and from season to season, resulting in significant variations of the analytical data.

The high figures given for the *Scenedesmus* sample from Germany demonstrate the close correlation between the polluted environment and contaminated algae because the plant from where this algal originated, was situated in close proximity to a heavy-industry complex.

Concentrations of dialkylnitrosamines of dimethyl- and dipentyl-nitrosopiperidine, nitrosomorholine, and nitrosopyrrolidine in samples of *Scenedesmus obliquus* from Peru and *Spirulina platensis* from India were below the detection limit of 0.1 μg kg^{-1}. At present, it is not possible to indicate an upper permissible concentration for the above chemicals because the health risk increases with increasing concentrations of the compounds. Furthermore, there is the probability that these chemicals are accumulated in animal or human tissue. For one of the most hazardous substances, benzo(a)pyren (formerly named 3,4-benz-pyrene), several countries specify an upper limit of 1 μg kg^{-1} in meat and meat products. According to the IUPAC recommendations, the level of this compound in single-cell proteins should not exceed 1 μg kg^{-1}. Because the amount of algae consumed daily will be far below that of meat, the amounts of the benzo(a)pyren detected in algal samples will be within the safety limit, if the recommendations for meat are applied to algal biomass.

Analyses on pesticide residues in algae have been carried out in very few instances only. Official regulations on permitted levels of pesticides or their residues in vegetables and meat products are in force in many countries. In addition, values for acceptable daily intake (ADI-values) for various

Table 13.10. *Analytical data of different algal samples for organic contaminants* ($\mu g\ kg^{-1}$)

	Scenedesmus (Germany)[a]	Scenedesmus (Thailand)[a]	Scenedesmus (Mexico)[b]	Scenedesmus (Israel)[c]	Chlorella (Israel)[c]	Micratinium (Israel)[c]	Spurilina (India)[d]	Spirulina (Mexico)[e]	Spirulina (Mexico)[f]
Fluoranthene	444.0	91.2	3.0	—	—	—	—	—	—
Benzo(a)pyrene	39.5	2.7	1.4	—	—	—	—	2.0–4.3	2.6–3.6
Benzo(b)fluoranthene	85.4	5.9	4.1	—	—	—	—	—	—
Benzo(k)fluoranthene	36.3	2.4	1.2	—	—	—	—	—	—
Benzo(ghi)perylene	52.1	3.9	1.4	—	—	—	—	—	—
α-HCH	—	—	0.1–0.03	—	—	—	18	—	—
β-HCH	—	—	—	—	—	—	3	—	—
γ-HCH	—	—	0.07–0.1	—	—	—	5	—	—
HCB	—	—	0.007–0.009	—	—	—	0.04	—	—
Dieldrin	—	—	0.03–0.08	—	trace	—	4	—	—
DDE	—	—	0.05	50	—	50	0.6	—	—
DDD	—	—	0.04	40	—	60	0.7	—	—
o,p'-DDT	—	—	—	40	—	—	—	—	—
DDT	—	—	—	80	—	30	6.0	—	—
PCB	—	—	—	700	600	—	0.5	—	—

Notes:
[a] Payer & Runkel (1978); [b] Becker & Venkatarman (1982); [c] Yannai *et al.* (1980); [d] Becker & Venkataraman (1984); [e] Boris & Tulliez (1975); [f] Durand-Chastel (1980).

pesticides are published in the WHO Technical Reports Series Nos 525, 545 and 574; these can be applied for algae too. As long as the algae are not treated directly with pesticides, the chances of excessive contaminations with these chemicals can be regarded as minimal.

Safety studies with animals

The toxicological safety of drum-dried *Scenedesmus obliquus* was tested by Pabst *et al.* (1978) by feeding this alga to mice. Over a period of seven generations mice of both sexes were given a diet containing 20% of algae for 12 weeks each; controls were fed an algae-free standard diet. Feed efficiency, body weight, reproduction, life span, organ weights and blood chemistry were recorded. In the algal-diet group, body weights increased by 10% compared with control, litter size was reduced by 11%, mean birth weight of pups increased by 4%. The weights of various organs in experimental and control animals differed as detailed in Table 13.11. The increase of weight of mice fed the algal diet was attributed by the authors to a higher feed intake, and the reduced litter size may have been caused by the higher body weight of the dams. According to the authors, the addition of 20% of algal powder in an otherwise balanced diet involves a considerable alteration in the crude fibre content and the proportion of the various nutrients.

This may have effected the alterations observed, especially the weights of the organs, which are linked closely to the metabolism. The only detectable difference in the haematological parameter was a 3% reduction of haemoglobin and haematocrit in algal-fed female mice. The litter size of the experimental animals was 11% smaller than that of the controls; the mean birth weight, however, was increased by 4%. All other parameters such as viability, gestation and lactation were comparable between the two groups. The females of the algal group displayed a highly significant increase in life expectance; after the period of 12 weeks the number of surviving animals was 48% higher than in the control group; no explanation was given for this prolonged life span. In a similar multi-generation study with rats, none of the deviating findings described above for mice could be reproduced, all parameters remained comparable between the experimental and control groups. It can be assumed that the differences observed with the mice are specific for this animal species; therefore they are not very suitable for experiments of this kind and have only seldom been employed in toxicological evaluations.

Feeding trials lasting 12 weeks with *Scenedesmus obliquus* on rats were also reported from India (Venkataraman *et al.*, 1977a, 1980). Drum-dried algae were incorporated into the diets at 20%, 15% and 10% protein levels; casein at 10% protein level was used as control. The food uptake was highest in the group containing 20% algae (1400 g) followed by the diets

Table 13.11. *Mean organ weights (g) of mice fed drum-dried* Scenedesmus acutus *and control diet at 10% protein level for 80 weeks*

Generation	Organ	Age (days)	Male Algae	Male Control	Female Algae	Female Control
F_4	Liver	10	2.39	1.94	1.93	1.31
F_3	Liver	40	3.50	2.79	2.29	1.61
F_5 and F_6	Liver	52	2.95	2.36	2.32	2.36
F_0–F_6	Liver	80	3.00	2.86	2.50	2.11
F_0–F_6	Brain	80	0.54	0.52	0.54	0.53
F_0–F_6	Kidney	80	0.75	0.71	0.50	0.44
F_0–F_6	Spleen	80	0.14	0.15	0.18	0.15
F_0–F_6	Testes	80	0.22	0.24	—	—

Source: Pabst *et al.* 1978.

with 15% (1300 g) and 10% (1000 g) algal protein and 10% (940 g) casein protein. This order was also found for the increases in body weight, which were 230, 205, 162 and 144 g, respectively. After the termination of the feeding period, various organs of the animals were examined: relative organ weights of liver, kidney, lung, thyroid and testes were greater in the control group. The calculated correlation between the weights of the different organs showed a close and normal relationship between the weights of liver, heart and kidney in all the groups. The livers of the rats fed the control diet showed mild centrilobular fat infiltrations, which could not be observed in the animals of the experimental groups. Analyses of hepatic enzymes showed a slightly increased activity of alanine aminotransferase in algal diets compared with the control, and a slight reduction in the activity of succinic dehydrogenase in rats fed on 15% algal protein.

In continuation of this investigation, differently processed *Scenedesmus obliquus*, namely drum-dried, sun-dried, sun-dried after previous cooking, was incorporated at 10% protein levels into the experimental diets in order to find out whether any of the drying methods cause toxic symptoms in the test animals during a feeding period of 12 weeks; casein was used as control protein. To illustrate the effect of these different diets on feed consumption and weight gain, the pattern of these data are summarized in Figs 13.6 and 13.7. The consumption of diets containing drum-dried algae was considerably higher compared with diets containing the sun-dried algal material. This difference seems to be caused by taste and smell preferences; the fishy smell of the sun-dried algae may be one of the contributory factors to explain the reason for this poor diet uptake. Rats fed with 20% drum-dried algae showed the highest weight gain; the animals fed sun-dried algae remained small and reached a maximum gain of only 35 g, which is shown quite impressively in Fig. 13.8.

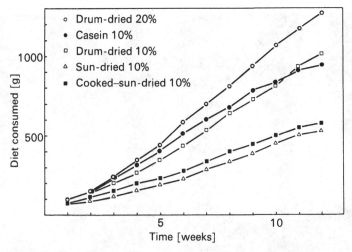

Fig. 13.6. Diet consumption by rats fed casein and differently processed *Scanedesmus obliquus*. Values are the mean of eight rats per group, feeding time 12 weeks.

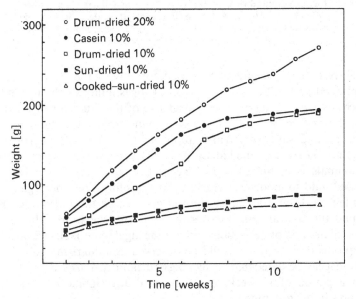

Fig. 13.7. Weight gain in rats fed casein and differently processed *Scenedesmus obliquus*. Values are the mean of eight rats per group, feeding time 12 weeks.

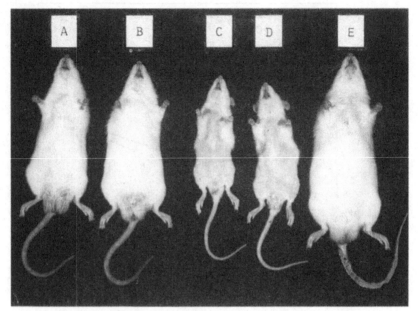

Fig. 13.8. Ventral views of rats fed on casein and differently processed *Scenedesmus obliquus*. A, Casein (10% protein); B, algae, drum-dried (10% protein); C, algae, sun-dried (10% protein); D, algae, sun-dried and cooked (10% protein); E, algae, drum-dried (20% protein).

These reduced weights were caused not only by the low feed intake as mentioned above, but also by the poor digestibility of the insufficiently processed algal material. The ventral side of the rats fed sun-dried algae were fairly free of hair. This loss of hair progressed from the center outwards and was observed after about eight weeks of feeding. The relative weights of heart, liver and kidney (g per 100 g body weight) were elevated in the animals fed sun-dried algae. Liver sections of rats fed on the casein diet showed mild to moderate centrilobular fat infiltration; the organs of the rats fed on drum-dried algae showed generalized fat infiltration, mainly around the central vein. Besides these effects, which are related to the physical method of processing and not the algal material itself, no adverse symptoms or indications could be observed. Summarizing all this information it can be stated that the feeding of drum-dried *Scenedesmus obliquus* over a period of 12 weeks did not reveal any significant toxicological features in the test animals.

Among the different algal species that have been selected for mass cultivation, the green alga *Micractinium* sp. has as yet gained little significance and consequently only very limited information is available on its possible toxic side effects. Preliminary toxicological tests with *Micractinium* sp. have been conducted in Israel, where this alga was cultivated on industrial effluents and harvested by alum flocculation. Yannai & Mokady (1985) incorporated drum-dried algal material at different levels into the rations for chickens, Japanese quails, and mice with the intention of establishing the safety of *Micractinium* for these animals as well as the safety of the chicken meat for humans. After seven weeks of feeding 7.5 and 15% of algae in the diet, no growth retardation or abnormalities of the chickens were found, indicating that the growth-promoting effect of the algae at these levels was comparable to that of the control diet. The quails were reared on diets containing 10 and 20% *Micractinium*. The only difference observed at the end of the four-week feeding period was a somewhat smaller number of eggs per female and a poorer hatchability; however, weight of eggs and one-day chickens was comparable to the control. As was noted for the chickens and the quails, all experimental groups of mice also exhibited a similar growth performance, no abnormalities were noted in the algae-fed mice, and their reproduction performance appeared to be normal.

More-detailed toxicological evaluations have been performed with *Spirulina*. The probably most comprehensive study was published from Mexico by Chamorro (1980). This report includes tests on sub-acute and chronic toxicity, reproduction, lactation, mutagenicity and teratogenicity. The author tested three different concentrations (10, 20 and 30%) of spray-dried *Spirulina*, produced in Mexico, which were added to a standard diet substituting the commonly used soybean meal. The amount of soybean meal in the ratios was reduced from 44% in the control to 30, 14 and 0%, respectively.

The total study, which lasted several years, was divided into several different experiments. In the first one, lasting for 13 weeks, weight gain of rats was monitored. No significant differences could be observed between controls kept on a standard diet and the experimental groups; the weight of the males varied between 352 and 340 g, and that of the females between 211 and 202 g. The relative weights of the various organs of all animals were comparable; detailed histopathological examinations did not reveal any anomalies and analyses of blood, urine, serum protein and some enzymes (GOT, GPT and AP) were within the normal range. A repetition of this evaluation lasting for 80 weeks gave similar results; no effect of the algal diets could be observed on the various parameters studied.

In order to detect any chronic toxicological effect of the algae, reproduction and lactation studies over three generations were conducted, lasting for two years. In all three generations fertility, gestation and lactation were recorded; the last generation was also subjected to a sub-acute toxicity study. No adverse effects were observed regarding fertility or litter size; the sub-acute toxicity study failed to reveal any harmful effect that might be caused by the consumption of the alga. In an additional mutagenicity test over three months, mice and rats were fed on diets containing 30% algae and then mated. The uterus of the females and the fetuses were examined for mutagenic effects but nothing abnormal could be detected.

In a teratogenicity study, which included rats, mice and hamsters, the animals were kept for various periods of time on the different algal diets mentioned. Fetuses were screened for internal and external malformations but, as before, no anomalies could be found.

The author came to the conclusion that none of the many parameters looked for showed any variation that might have been caused by the feeding of the three algal diets; differences observed were not based on dose effects but were caused by external factors and could not be reproduced.

In another study, reported by Boudene *et al.* (1975), *Spirulina* was fed to one generation of rats in rations containing 25% algae over a period of 75 weeks. This evaluation also failed to demonstrate any toxic effect of this alga.

These encouraging results were confirmed by other authors. Even the feeding of *Spirulina* as the sole source of protein to sexually maturing rats, as was done by Contreras *et al.* (1979), did not provoke any unusual observations of body weights, organ weights, histology, haematology, levels of gonadotrophic, luteinizing and follicle-stimulating hormones or prolactin. No apparent signs of organ or body toxicity were observed. The effect of feeding sun-dried *Spirulina* (10% protein level) for 12 weeks as the sole source of protein was also examined by Becker & Venkataraman (1984). The consumption of the algal diet was less compared with the casein control and so was the weight gain. *Spirulina* as the sole source of protein does not effect the breeding performance of rats, as evidenced by Chung *et al.* (1978) and demonstrated by the normal conception, litter production and lactation of the animals. Histological observations of different organs of males and females showed no abnormalities. The effect of feeding two *Spirulina*-rich diets (730 and 260 g algae per kg diet) for 100 days to rats was studied by Bourges *et al.* (1971) After 50 days of feeding the two diets and a casein control to different groups of animals, the groups were switched either from control to algal diet or vice versa in order to detect possible age differences in response. In spite of the high algal content in both algal diets all animals survived with apparent health, all organs were normal macroscopically and microscopically and no differences were found between control and experimental groups.

These encouraging findings on the safety of *Spirulina* as a protein substitute were confirmed by additional long-term feeding studies and tests for chronic toxication (Bizzi *et al.*, 1980; Yoshino *et al.*, 1980). In spite of very extensive biochemical evaluations, haematological tests and histological examinations of several organs, no abnormalities or differences between experimental and control groups could be found.

To round up the spectrum of toxicological evaluations, the tests on possible dermal toxicity should be mentioned, which were performed by

Krishnakumari *et al.* (1981) with drum-dried *Scenedesmus obliquus* and sun-dried *Spirulina platensis*. For the investigation on allergic affects of the algae, different dosages of both algae were applied on clipped skin on the dorsal side of rats. Within the two-week observations period, none of the animals showed any sign of erythema or oedema on the skin and hair growth was resumed as in controls.

Summing up all the available information on possible toxic properties of the different algae tested so far, it can be stated that none of these species showed any negative effects in animal-feeding experiments. Rodents, employed for these investigations, accepted the algal-containing rations very well, even at high algal concentrations, which in several cases resulted in increased feed consumption and weight gain. No serious anomalies were found, neither after feeding algae for shorter periods of several months, nor after long-term studies lasting for two–three years. Multi-generation studies demonstrated that the algae had no effect on reproduction or the performance of offspring. Studies on acute or chronic toxicity failed to reveal any evidence that would restrict the utilization of properly processed algal material, and it is quite unlikely that further studies will disclose any toxic potential of the algae described here.

13.8. Animal feed

The great expectations on the immediate use of algae as human food have subsided for various reasons. Today, it seems that the utilization of algae as animal feed will become a more distinct possibility in the near future. The large number of nutritional and toxicological evaluations that were conducted in the past demonstrated the suitability of algae biomass as a valuable feed supplement. Besides these basic studies, a series of detailed feeding trials were performed with particular target animal species. These trials were designed especially in order to find out the maximum amount of algae that could be incorporated in feed formulations as a substitute for conventional protein sources (soybean meal, fish meal, rice bran) without causing negative effects. Because most of the algae in question, with the exception of *Spirulina*, contain considerable amounts of cellulosic cell-wall material, which may cause problems in digestibility, one must distinguish between two groups of animals in this context: ruminants and non-ruminants (monogastric vertebrates). Ruminants such as sheep and cattle are capable of digesting cellulosic material through the action of cellulase synthesized and released by symbiotic bacteria in the rumen. Hence, it should be possible to feed native algae directly to ruminants; however, this possibility has not gained much attraction yet. Monogastria, on the other hand, including humans, are not able to digest cellulosic plant material so that algal biomass has to be processed properly before it is suitable as feed.

Table 13.12. *Performance of poultry fed on diets containing various concentrations of different algae*

Diet	Weight gain (g)	FER[a]	Reference
Control (4 weeks)	637	1.67	
Chlorella/Euglena, 25% replacement of soya protein in the diet	646	1.67	
Chlorella/Euglena, 50% replacement	637	1.61	
Micractinium, 25% replacement	629	1.70	Mokaday *et al.*
Micractinium, 50% replacement	600	1.65	(1980)
Oocystis, 25% replacement	543	1.78	
Oocystis, 50% replacement	571	1.81	
Scenedesmus, 25% replacement	644	1.71	
Scenedesmus, 50% replacement	615	1.77	
Control (4 weeks)	135	3.10	
Control + 10% *Chlorella* (vacuum-dried)	262	2.40	Combs (1952)
Control + 10% *Chlorella* + 0.1% met	298	2.30	
Complete broiler mash	342	2.20	
Scenedesmus (24% protein) + 0.1% met	—	5.20	
Scenedesmus (16% protein) + 0.1% met	—	4.40	
Scenedesmus (2.9% protein) + soya (21.1% protein)	—	2.30	
Scenedesmus (4.8% protein) + soya (19.2% protein)	—	1.90	Brune & Walz (1978)
Scenedesmus (4.8% protein) + fishmeal (2% protein) + soya (17.2% protein)	—	2.30	
Control (18% soya protein) + 0.54% met (3 weeks)	237	1.83	
Scenedesmus/Chlorella (18% protein)	87	3.58	Leveille *et al.* (1962)
Chlorella (18% protein)	9	17.90	
Spongicoccum (18% protein)	13	17.00	
Control (4 weeks)	667	3.27	
Scenedesmus (4% of diet)	4.61	4.17	
Scenedesmus (7.5% of diet)	385	4.95	Lincoln & Hill
Scenedesmus (10% of diet)	374	5.07	(1980)
Spirulina (5% of diet)	636	4.53	
Spirulina ((10% of diet)	466	5.08	
Control (22% protein) (3 weeks)	484	1.60	
Chlorella (7.5% of diet, 22% protein)	485	1.64	Lipstein & Hurwitz
Chlorella (15% of diet, 22% protein)	474	1.63	(1980)
Control (19% protein) (8 weeks)	1878	2.20	
Chlorella (15% of diet, 19% protein)	1732	2.29	
Control (20% protein) (18 days)	383	2.25	
Spirulina (19% protein)	384	1.92	
Spirulina (19% protein), extracted with methylene chloride	402	2.01	Brune (1982)
Scenedesmus (19% protein)	282	2.61	
Scenedesmus (19% protein), extracted with methylene chloride	301	2.36	

Table 13.12. (*contd.*)

Diet	Weight gain (g)	FER[a]	Reference
Control (21% protein) (3 weeks)	173	1.99	
Spirulina (5% of diet)	167	2.03	
Spirulina (10% of diet)	164	2.06	
Spirulina (15% of diet)	165	1.97	
Spirulina (20% of diet)	146	2.04	Ross & Dominy
Control (22% protein) (6 weeks)	1674	1.78	(1989)
Spirulina (1.5% of diet)	1683	1.78	
Spirulina (3% of diet)	1706	1.80	
Spirulina (6% of diet)	1692	1.78	
Spirulina (12% of diet)	1610	1.77	

Note:
[a] FER = feed efficiency ratio = feed consumed (g) per weight gain (g).

Poultry

Most of the trials on animals have been conducted on poultry and it can be assumed that the incorporation of algae in poultry rations offers the most promising prospect for their commercial use in animal feeding. The capability of poultry to excrete uric acid allows elevated algal concentration in the diet without endangering the animals.

Selected data of respective investigations are compiled in Table 13.12. A few evaluations shall now be described briefly. The tests performed in Israel with sewage-grown algae with the aim of replacing soybean meal used in broiler mash (Mokady *et al.*, 1980) indicate that all algal species tested can successfully replace 25% of the sóya protein (i.e. 5% algae); higher amounts of algae (15%), however, lowered the feed conversion efficiency. The algae could not replace fishmeal as a source of growth factors in these studies.

A similar trend was observed by incorporating *Scenedesmus obliquus* as a replacement for groundnut meal (8%) and fishmeal, and by partially and completely replacing fishmeal by *Scenedesmus* sp. and *Spirulina* sp. In all these studies, reduced growth was observed in the groups fed containing algal diets. Based on these findings it was recommended to use algal concentrations not higher than 5% in the feed preparations.

These findings are in contrast to the observations of Reddy *et al.* (1978) who tested the performance of chickens fed *Scenedesmus* sp. at the 5% protein level in place of fishmeal and who came to the conclusion that the algal protein can be used in poultry rations as supplement for fishmeal. These divergent results might be caused by differing feed preparations, which in the latter case have compensated for deficiencies.

The nutritive value of a new strain of *Chlorella* (*C. vulgaris A1-25*) was evaluated in Japan in poultry-feeding experiments (Yoshida & Hoshii, 1982). This alga has a very thin cell wall and it was assumed that it would be more easily digestible than chlorophyceae in general. The alga contained less protein (24%) but more carbohydrates (53% nitrogen-free extract) than the common strains of *Chlorella*. Air-dried samples were incorporated at 16.3 and 20% levels in poultry rations. Gross protein value with supplemental methionine (2%) was 71%. The algal meal was well accepted and no signs of acute toxicity could be detected in the birds.

It remains to be seen whether isolation and breeding of cell-wall-free chlorophyceae or at least the selection of strains with fragile cell walls offer new possibilities to overcome the problems related to processing and digestibility of *Chlorella* or related green algae species.

Feeding studies with different concentrations of *Spirulina* (up to 30%) showed that both protein and energy efficiency of this alga were similar to other conventional protein carriers up to a level of 10%, but were reduced at higher concentrations. In broiler experiments with *Spirulina* sp., weight increase was depressed from the first day that the alga completely replaced traditional proteins in the diet (Blum & Calet, 1975). The reduction in growth was less with algal concentrations up to 5% in the diet; increase in weight, however, was reduced by 16 and 26% at algal levels of 20 and 30%, respectively.

Only limited information is available on the nutritional value of algae other than *Chlorella*, *Scenedesmus* and *Spirulina*. One of the few poultry-feeding studies on uncommon algal species was carried out by Lincoln & Hill (1980) using sewage-grown *Euglena* sp. and *Synechocystis* sp. as ingredients in broiler rations. No specific toxicity was noted in the experiments; however, the authors indicated that *Synechocystis* may be toxic under certain culture condition. Concentrations up to 10% gave good results, whereas inclusion of 15 and 20% algae reduced the growth of the birds.

In all the studies described above, whole algal biomass was added to the diets and only very few investigations were carried out to study particularly the effects of isolated algal ingredients.

In order to elucidate the nutritive value of purified algal protein without possible inhibiting or stimulating properties of any other constituents, fractions of *Spirulina* sp. and *Scenedesmus obliquus* (extraction of lipids by methylene chloride) were used as the sole source of protein by Brune (1982). As can be seen, the lipid-free material gave similar or even better results than the untreated algae. Contrary to other reports, in this case it was possible to feed the extracted algal material as the sole source of protein without causing growth retardation in the animals.

These findings suggest the possibility that growth-depressing effects of untreated algae could be caused by a high lipid content, especially long-chained polyunsaturated fatty acids, in the algal biomass.

Because the algae are harvested by aluminum-induced flocculation at various places, the effect of dietary aluminum on chicken was tested by some authors. However, the data generated are not yet conclusive. Although some authors reported that aluminum (8% in the algal biomass) caused significant growth reductions that could not be influenced even by acid extraction of the algal biomass, others did not observe any growth depression of chickens fed aluminum-containing diets.

There are a few reports in the literature on the effect of algae on the performance of laying hens. For sewage-grown *Chlorella* sp., no differences were found in egg-production rate, egg weight or food conversion efficiency between controls and birds receiving up to 12% of this alga. In order to evaluate the effect of sun-dried *Spirulina platensis* as an ingredient in starter and layer rations, feed conversion efficiency, egg-laying properties and egg-quality criteria were examined in a detailed study reported from India. Three experimental diets were prepared using the alga to replace either groundnut cake or fishmeal or both simultaneously on isoproteinic basis. For starter feeds, the protein content was about 21%; after eight weeks the layers were supplied with a feed containing 15% protein. Total duration of the experiment was 64 weeks with the egg-laying period lasting for 44 weeks. The performance of the layers kept on a control diet and the birds kept on algal-based diets was comparable with feed conversion ratios ranging between 2.41 and 2.64. Egg laying commenced from the twentieth week and the egg weight slowly increased with the age of the birds. There was no significant difference in the egg quality between the groups; size, weight, shell thickness, solid content of the egg, albumin index etc. were comparable. The yolk, however, had a distinct intense orange color in layers fed the algal diet.

·   Other studies on effects of partial substitution of soybean meal in layer rations by *Spirulina* showed that at algal concentrations above 10% the sensory characteristics of the eggs were affected, resulting in a 'chemical flavor' of the egg.

The yellow color of broiler skin and shanks, as well as of the egg yolk, are the most important characteristics that can be influenced by feeding algae. This pronounced pigmentation is primarily caused by the algal carotenoids consumed in the feed and deposited in the fat of the animals. Natural pigments are currently obtained from corn gluten, alfalfa meals and marigold petal extracts.

In several countries, poultry rations have traditionally been supplemented with plant-pigment concentrates as the supply of pigments to satisfy consumer preferences for colored egg yolks and broiler carcasses. In these cases algae would be an optimal source of pigments besides their property as a valuable protein supplement. Lutein and zeaxanthin are about equally effective as coloring agents, the cryptoxanthanes only about as half as effective whereas neoxanthin or violaxanthin are ineffective in chicken or egg pigmentation. Lutein produces a more yellow color whereas canthaxanthin results in red coloration. However, it has to be kept in mind that the pigmentation may reach unacceptably high levels or may impair the quality of the broiler in such areas where non-pigmented meat is preferred. Summarizing the available information it can be stated that most of the authors cited here come to the conclusion that in poultry rations algae up to a level of 5–10% can be used safely as a partial replacement of conventional proteins. Higher concentrations, however, cause adverse effects on prolonged feeding. Algae are, in general, officially approved in several countries as chicken feed and do not require new testing or approval. However, it has to be decided from case to case how restrictive the different algae species are regarded as feed supplements.

Pigs

The possibility of using algae as a part of pig feed preparations has been tested in only a few places. Some of the major findings shall be given here to elucidate the potential of algal biomass as a substitute or supplement in commercial pig mash.

A mixture of *Chlorella* sp. and *Scenedesmus* sp., grown on sewage and dried either in air or by drum-drying, was incorporated at 2.5, 5.0 and 10% levels in pig rations, substituting soybean and cotton-seed meal (Hintz & Heitmann, 1967). The feed efficiency ratios estimated after a feeding period of seven weeks were 3.76, 3.85 and 3.90, respectively, and almost identical to the control value of 3.85. In additional investigations, the authors tested the supplementary effect of oven-dried (70–80 °C) *Scenedesmus* in three different feed preparations, i.e. group I containing only fishmeal as protein, group II composed of 75% algal and 25% fishmeal protein and group III consisting of 75% soya and 25% fishmeal protein. During the growth period between 40 and 110 kg body weight, the average daily feed consumption was highest in group II, the differences from the other groups being highly significant. In spite of the slightly increased daily gain in weight, the feed conversion efficiency in group II (3.59) was higher compared with group I (3.33) and group III (3.44). In further studies, the influence of fishmeal, soybean meal and drum-dried *Scenedesmus* on the growth and the quality of the pork was tested. With respect to growth, no significant differences could be found between the different groups consisting of different protein sources (group I, fishmeal; group II, soybean meal; group III, 75% soybean meal and 25% fishmeal; group IV, 25% soybean and 75% algae). Fishmeal and the mixture of algae plus soybean meal had a significantly positive influence on the ripening process of the meat.

Substituting alfalfa leaf meal with drum-dried *Scendesmus* and casein as standard on an iso-nitrogeneous basis resulted in comparable growth performances. Both algae as well as casein gave relatively high apparent digestibility (75% and 100%) as compared to alfalfa (60%). Algae fed at a

12.5% level was found to be acceptable and highly compatible, resulting in good biological value and protein utilization. Some investigators even concluded that *Scenedesmus* protein is extremely suitable to be utilized as the *only* protein supplement for fattening pigs.

This assumption, however, could not be confirmed by others, because it was observed that the addition of higher concentrations of algae in pig rations led to reduced digestibility and growth depression. For example, feeding tests carried out at the algal project at Singapore showed that replacement of 50% of the soybean meal, given at 15.5% in the diet, with steam-boiled sewage-grown algae resulted in acceptable values; total replacement, however, produced significant reduction in both the weight gain and the feed efficiency ratio, thereby demonstrating that the nutritive value of the algae was clearly inferior to that of the soybean meal.

Besides *Chlorella* and *Scenedesmus*, very few trials were run on the use of *Spirulina* in long-term feeding studies on pigs. One of the few investigations that employed this cyanobacterium as a feed additive was conducted by Fevrier & Sevet (1975). A first series of experiments was performed on early-weaned piglets (from 12 to 42 days) with an algal supplement representing 12% of the total proteins replacing skim milk or soybean meal. The apparent digestibility decreased slightly from 84.0 for the control diet to 77.9% for the algal diet. In a second study, *Spirulina* was fed continuously at 5% to sows during the first two reproductive cycles. At the commencement of the investigation the characteristics of the experimental and the control group were identical and the reproductivity of the sows of the two groups was the same. Although the feeding of *Spirulina* did not seem to give rise to any problem, the authors recommended that incorporation of algae should be restricted to a level not exceeding 25% of the total dietary protein, especially in the young animals, because elevated amounts of algae may lead to an insufficient supply of digestible substances thus reducing growth and feed efficiency.

More recently, two studies were reported on the feasibility of replacing 33% of soy protein in a basal diet with proteins from the two cyanobacteria *Spirulina maxima* and *Arthrospira platensis* (the taxonomical classification of this species is questionable) and *Chlorella* sp. to pigs four to eight days old weaned to a dry diet (Yap *et al.*, 1982). Animals fed the basal diet up to the twenty sixth day gained weight at a rate not significantly different from those fed alga diets. There was no sign of diarrhea, loss of appetite, toxicity or of gross histopathological lesions of the gastro-intestinal tract. The authors suggest that at least 50% of the protein supplied by soybean meal (33% of total) can be replaced by these algae without adverse effects.

Summarizing the available information it can be concluded that nearly all the animal feeding studies come to the result that microalgal biomass is a feed ingredient of good nutritional quality and suited very well for rearing pigs, provided the material is processed properly. It can replace conventional proteins like soybean meal or fishmeal and no difficulties in acceptability of algae were reported for these animals. Another promising factor enhancing growth of pigs and other animals has been speculated in some reports, where it is assumed that the algae have some antibacterial properties that are not available in standard rations. However, how far this idea is true or not remains open.

Ruminants

The possibility of using wet algal cakes or similar preparations, which can be sterilized suitably without further processing before feeding it to cattle or other ruminants, has not received much attention yet, although it would be the most simple method of using algal biomass as animal feed. One of the reasons for the very limited number of trials is obviously the large amount of algae required to perform appropriate feeding experiments with these animal species.

One of the very few studies on the utilization of fresh untreated algae for feeding ruminants has been reported from Bulgaria (Ganowski, Usunova & Karabashev, 1975). Here, 1 l of concentrated native *Scenedesmus obliquus* ($2–3 \times 10^8$ cells ml^{-1}) were fed to calves over a period of three weeks. It was observed that this feeding increased the contribution of the intestine in the digestive process without facilitating the digestion of the feed. Only minor differences were observed for the digestibility between control and experimental animals.

In their very detailed studies, Hintz *et al.* (1966) have evaluated the nutritive value of sewage-grown algae on lambs, sheep and cattle. Differently processed (air-dried and drum-dried) mixtures of *Chlorella* sp., *Scenedesmus obliquus* and *S. quadricauda* were fed to wethers. The rations consisted of 60% algae and 40% hay; alfalfa/oat/hay diets served as control. The protein digestibility of all diets was about 70%, the digestibility of the non-lipid, non-protein organic matter ('carbohydrate'), however, was much lower in the algal diet than in the control, which can be attributed to the crude fiber content of the algae. For weaning lambs, cotton-seed meal, alfalfa pellets and alfalfa/algae (10:4) pellets were compared. The alfalfa pellets were non-satisfactory because the animals could not consume that amount of feed required to supply the nitrogen needed for maintenance. With sufficient amounts of the other two diets the lambs were able to maintain their weight or even to gain weight.

In a further study with drum-dried algae, beef steers were used as experimental animals. The rations tested were composed of alfalfa/hay, algae/hay (2:8) and algae/hay (4:6). Because the consumption of the ration containing the higher amount of algae was poor, intake of the other diets was limited to the daily intake of the algal ration. The addition of algae at both levels did not decrease the digestibility of the crude protein (74%), but the feed with the higher algal content showed a reduction of carbohydrate digestibility from 68 to 52%, similar to that obtained in the trial with sheep.

Although in India ruminants are not considered as the primary target group whose diets have to be supplemented with algae, some preliminary feeding studies have been performed there on the usefulness of *Spirulina* sp. as a replacement for groundnut cake (30%) in the rations for infant calves. After a feeding period of 28 weeks, the weight gain of the experimental animals was reduced by 8% only compared with the control group, indicating that this alga is a suitable substitute for protein-rich oil cakes in the concentrate mixture of calf rations.

The value of an uncommon feed mixture composed of algae and paper was tested on sheep by Davis, Sharkey & Williams, 1975. After harvesting sewage-grown algae on paper filters, the filters plus algae were mixed with barley (3:7) into the rations and compared with other rations of fresh paper/barley (3:7) and alfalfa/barley (3:7). The animals utilized the algal/paper mixture poorly. Its incorporation as roughage in a high-grain diet ensured its consumption by the sheep but intake was reduced and they digested less energy and nitrogen than could be achieved with an alternative source of roughage such as alfalfa.

Sewage-grown *Chlorella* sp., harvested by alum flocculation and dried by drum-drying was fed to young rams. The algae substituted 50% of the dietary protein of a diet based on soybean meal as the standard protein carrier. It was found that the digestibility of the dry matter of the algal diet was reduced to about 70% compared with the 80% of the control diet. Apparent nitrogen digestibility of the experimental diet was 71.3% whereas the value for the control diet was 83%. The reason for these differences might be the high amount (25.5%) of non-absorbable minerals in the algae. Among these minerals, about 6% was aluminum remaining from the harvesting. Because it was expected that this high aluminum concentration would interfere with phosphorus absorption, dicalcium phosphate was added to the diets. In ruminants, there is an intensive endogenous phosphorus secretion into the stomach that might be much higher than the amount of ingested phosphorus, so that elevated amounts of aluminum ingested with the feed would cause a depletion of body phosphorus reserves. These considerations indicate that for ruminants the feeding of aluminum-flocculated algae might not be very suitable as it may lead to phosporus deficiencies.

13.9.   Nutritional studies with humans

In the previous section we saw that detailed toxicological studies on animals failed to disclose any hazardous or antinutritional properties of the microalgae tested. The results permit the conclusion that appropriate amounts of these algae can be utilized as animal feed without taking the risk of endangering the health of the animals. However, before microalgae can be declared as a safe protein supplement for human consumption, further pre-clinical and clinical tests are required. Unfortunately, only a limited number of nutritional and toxicological studies on humans have been conducted so far, so the available information is still too scarce and incomplete for a concluding statement. One point, which has caused serious concern on the utilization of algae for human consumption, is the elevated amount of nucleic acids found in algae. The toxicological and clinical relevance of this aspect has already been discussed.

In this section, more general questions shall be touched on that are related to the use of algae as food or food supplements. No systematic testing program has been developed yet and the studies performed cover quite diversified possibilities of applied algology. The objectives aimed at by the different investigations range from studies with malnourished infants up to mass feeding trials in government-subsidized kitchens.

It is striking that the majority of tests with humans were performed before 1970. There appear to be several reasons for this. Firstly, the alarming reports on global protein shortages published by international organizations during the 1960s gave rise to the development of several projects working on the production of unconventional sources of protein. In connection with these schemes and supported by the idea of using algae as a proteinaceous food stuff on long space flights, various tests with volunteers were carried out. Secondly, in those days there was a lack of awareness of possible harmful effects that might be caused by the consumption of algae, and neither guidelines nor recommendations were available that would have stipulated the performance of nutritional and toxicological tests, especially with humans. Last, but not least, criticism during the past decade on the exploitation of unconventional sources of protein in general has affected the efforts relating to the use of algae as food so that the need for immediate trials on humans has subsided.

It is obvious that even in the scientific studies contradictory results were obtained about the reaction of humans to algal diets. There are reports stating that people have lived solely on algae for certain periods of time without developing any negative symptoms, whereas in other studies discomfort, vomiting, nausea and poor digestibility of even small amounts of algae were among the symptoms reported.

In the following, only those studies will be reviewed that were conducted with the intention of elucidating the nutritional value of algae used as food for humans and also those that were designed with the intention of detecting possible risks connected with the utilization of algal biomass. We begin by casting a retrospective glance on the earlier investigations. It has to be stated that several of these tests would not be performed nowadays because insufficient precautions were taken over the health of the persons being tested. In addition, it has to be assumed that the algal material used in the early investigations was not processed properly, mainly because the importance of post-harvesting treatment was not recognized fully at that time.

In one of the first studies (Powell, Nevels & McDowell, 1961), a mixture of *Chlorella* sp. and *Scenedesmus* sp., cooked for 2 min and dried under vaccum, was given to five healthy, 18–23-year-old men (in a later review it was mentioned by the same authors that the algae were boiled for 2 h in an autoclave, methods that are both unsuitable for the processing of these algae). During different experimental periods, which lasted between three and seven days, the individuals received diets supplemented with algae in such a way that the daily algal consumption per head amounted to 0, 10, 20, 50, 100, 200 and 500 g. All men tolerated amounts up to 100 g; at all algal concentrations the taste was reported to be strong and disagreeable. Abdominal distention associated with increased eructation and flatulence was noted at the beginning of the study and became more serious at the 200 g algal level, connected with nausea, vomiting and hard stools. Only two people completed the period with 500 g of algae per day, but complained of abdominal cramps and malaise. Beside these symptoms associated with gastro-intestinal disorders, no other anomalies were reported by the authors based on laboratory tests and physical observations. No doubt the adverse effects were mainly caused by the insufficient processing of the algal material that left most of the biomass indigestible for the test individuals. In addition, it has also to be mentioned in this context that the administration of such high amounts of algae unnecessarily endangered the health of the test persons without producing any conclusive results, mainly because the duration of the test period was too short. The duration of such studies is of general importance, especially in view of the findings reported some years later by other authors (Kofranyi & Jekat, 1967) who determined the biological value of different proteins including algae alone and in mixtures with egg. According to these authors the experimental period for each protein mixture to be tested has to last for at least three to four weeks, because it takes 9–11 days until an equilibrium has been established between N-intake and N-excretion. The minimum requirements to achieve this balance were 0.5 g of egg, 0.57 g of milk, 0.62 g of drum-dried *Scenedesmus obliquus* or 0.8 g of wheat per kg of body weight per day, whereby a mixture of egg and algae (4:6) was superior to egg alone. These experiments were carried out over a period of six months without any complications or signs of discomfort.

In 1965, another study was conducted with five young men (aged 27–33 years) who were given *Scenedesmus obliquus* (lyophilized and autoclaved for 30 min) as the sole source of nitrogen at a rate of 7.1 g N per day, which was about 45 g of protein or 100 g of algae (Dam *et al.*, 1965). The men were able to ingest the algae but also complained about upset stomach and nausea.

Probably in view of the possibility of using algae as a protein supplement in space flights, Russian scientists tested the effect of three weeks' uptake of diets containing 50, 100 and 150 g of freeze-dried algae (a mixture of *Chlorella* sp. and *Scenedesmus* sp.) (Kondratiev *et al.*, 1966). The authors analyzed different blood parameters, urine and feces of the test individuals. For the groups receiving 50 and 100 g of algae daily, only negligible changes of the metabolic parameters were observed, and, except for some indices characteristic for lipid metabolism, all analytical data remained within the limits of the physiological standard. The daily uptake of 150 g of algae, however, led to some shifts in the state of health of the majority of the test persons that led to the conclusion that the maximum amount for daily consumption of algae is 100 g without considering the limits imposed by the nucleic-acid content. This assumption was confirmed by studies performed in Germany a few years later (Müller-Wecker and Kofranyi, 1973). Test persons were given a diet, which initially was composed of 25% drum-dried *Scenedesmus obliquus* and 75% whole-egg protein, over a period of four weeks during which the proportion of algae was increased stepwise until the algae served as the sole source of protein (77 g per day) for another four weeks. No negative effects were observed during the complete experimental period; the biological value of the algae differed between 81.5 and 96 compared with egg protein, indicating an optimal processing of the material used. Mixtures of algae with other conventional proteins (fish, flour, soybean flour, egg, rice and gelantine) were also tested at a daily nitrogen level of 6 g. All preparations supported positive nitrogen balance except the diets containing algae alone or the combination algae/gelantine. The value for the rice/algae mixture was similar to those obtained for other high-quality proteins (soybean, egg), demonstrating the beneficial effect of algae as a source of lysine and threonine.

A different attempt at testing algae-containing diets was reported from Peru (Gross *et al.*, 1978),. where young naval cadets and school children were employed to evaluate their tolerance to drum-dried *Scenedesmus obliquus* as a food ingredient. Hematological data, urine, serum protein, uric-acid concentration, and weight changes were measured. During the four-week test period, the adults received 10 g and the children 5 g of algae daily, incorporated into their normal diet. No changes in the analyzed parameter were found except a slight increase in weight.

There was a unique attempt by the same authors to test the value of this alga as a food supplement in the diets of hospitalized slightly (group I) and seriously undernourished (group II) children. The four-year-old children of group I received 10 g of algae daily over a period of three weeks. They showed a significant increase in weight (27 g per day) compared with other children of the same group who received a normal diet; no adverse symptoms were recorded. The second group was nourished with a diet enriched with 0.87 g of algae per kg body weight, substituting for 8% of the total protein only. In spite of the low contribution of algal protein, the daily increase in weight was about sevenfold over diets without algae and all anthropogenic parameters were shifted towards normal at the end of the study. This positive effect has to be attributed to certain therapeutic properties of the algal biomass, because diarrhea was cured during the algal uptake so that a better resorption of the nutrients by an activation of epithel cell regeneration in the gut can be assumed.

Finally, nutritional studies that were performed with volunteers in Thailand within the framework of a Thai–German algal project should be mentioned. For three weeks, the test individuals were offered a diet consisting of 50% of algal protein (drum-dried *Scenedesmus obliquus*) and 50% of vegetable protein. An increase in the uric-acid concentration of 2–3 mg 100 ml^{-1} was observed in the persons taking the algal diet so that the diet composition had to be modified. The digestibility of the algal protein was good; however, the excretion of nitrogen was higher compared with egg protein.

13.10.   Quality standards

Commercially produced microalgae, or their products, are already being marketed and accepted to a certain degree in some industrialized countries, where consumers are responsive to quality information given with the product and where education on the nutritional properties of algae helps to overcome certain prejudices.

Nevertheless, the introduction of new foodstuffs or food supplements such as microalgae into the conventional diet of developed countries is still difficult because of inexperience with this unconventional food item. It is even more afflicted with various complications in developing countries, because it often affects conservative ethnic factors, including religious and socio-economic aspects.

The resulting reluctance against microalgae among the potential consumers is increased by public-health difficulties concerning the harmlessness of microalgae and, with very few exceptions, the lack of official legislative regulations on the production and composition of algae-based foodstuffs.

To overcome these obstacles, immediate action is required by scientists, nutritionists and administration. However, in the case of algae, which are still considered as novel, unconventional foodstuffs, it is necessary to evaluate carefully with appropriate responsibility whether the complete internationally recommended testing program for single-cell protein has to be applied or if faster and simpler methods are available that can establish the safety of this unconventional foodstuff. The fulfilment of the present PAG guidelines is difficult or impossible for financial and/or technical reasons, especially in developing countries that could otherwise be interested in the production of algae or other sources of single cell protein.

The necessary recommendations about testing and utilization of algae should probably include the following specifications.

1. Details on technical properties in order to assure uniformity of the product, so that algae, produced in any country, are of equal quality to the product that was used in fulfilling the quality criteria. Information on production, processing and storage of algae and on possibilities of supplementation with regular diets should be given.
2. Details on safety regulations, in order to assure freedom from hazardous physical or chemical substances, from toxicological effects or pathogenic microorganisms, which are harmful to humans as well as to animals.
3. Data on nutritional value, mainly with regard to protein and essential amino-acid content, digestibility and protein utilization.
4. Information on compatability, with respect to physiological, nutritional and esthetic factors.

As far as is known, India seems to be the only country that has so far stipulated a legislative standard for food-grade *Spirulina* through the Ministry of Health. These official requirements, as well as suggested guidelines for quality of microalgae given by Jassby (1988) and data based on international guidelines and on his own experiments and results collected from various sources, are summarized in Table 13.13 as a proposal for quality criteria of algal products. This recommendation, of course, is not obligatory but rather a suggestion for fruitful discussion.

The facts that almost no adverse symptoms have been revealed so far and that side effects caused by algae appear to be extremely rare should be compared realistically with the grave problems of malnutrition and feed or food shortages in several countries. As long as specific regulations are missing, the available amendment 'Good Manufacturing Practice' (GMP) (1985) can be applied that covers all aspects of food processing and has also been adopted for specific foods although microalgae are not included. In addition to GMP, national regulations for milk products and infant food can be applied, which mostly cover hygienic, microbial and toxic aspects.

Table 13.13. *Quality criteria for algae*

	A	B	C
Crude protein (%)	>45.0	55	
Total ash (%)	<10.0	9	
Acid insoluble ash (%)		0.5	
Moisture (%)	<10.0	9	<7
Nucleic acid (%)	<6.0		
Lead (mg kg^{-1})	<5.0	2.5	<2.0
Arsenic (mg kg^{-1})	<2.0	1.1	<2.0
Cadmium (mg kg^{-1})	<1.0	1.0	<0.1
Mercury (mg kg^{-1})	<0.1	0.1	<0.1
Pheophorbide			<1.2 mg g^{-1}
Coliform bacterial count	absent in 0.1 g		negative
Salmonella sp.	absent in 1 g		negative
Shigella	absent in 1 g		negative
E. coli	absent in 1 g		negative
Mold			<100 g^{-1}
Filth (insect fragments)			<300 per 100 g
Rodent hairs			<1.0 per 100 g
Pesticides	According to the directions of respective food laws, PAG and IUPAC guidelines		
Vitamins and mineral requirements (100 g^{-1})			
Vitamins of the B complex group (mg)		6	
Vitamin B$_{12}$ (μg)		100	
β-Carotene (mg)		160	
Phosporus (mg)		850	
Iron (mg)		100	
Zinc (mg)		35	
Potassium (mg)		1350	
Nutritive criteria			
Protein efficiency ratio	>1.80		
Biological value	>75		
Digestibility coefficient	>75		
Net protein utilization	>55		

Notes:
A, General recommendations.
B, Indian Standard IS 12895 : 1990, Amendment No. 1, June 1991 for *Spirulina*, food grade-specification.
C, Suggested guidelines for quality of microalgae powder (after Jassby, 1988).

14   Applications of algae

The general objective of microalgal biotechnology is to produce specific products that meet the need of the consumer. This ranges from the defined utilization of the whole algal biomass as a source of protein or the extraction of certain valuable constituents (pigments, enzymes) to the sometimes difficult to verify application as therapeutic agents and the nebulous utilization as a health food. The last application does not constitute a concrete product as its demand has been created without any need or efficacy. Although the health-food market has seen considerable ups and downs depending on actual consumer trends and the fantasy of the algae-promoting companies in attributing beneficial properties to the consumption of algae, it is still the branch of algae production with the highest sales. In recent years, processes for the isolation of a number of high-value constituent of algae have been developed, including different pigments (β-carotene, phycobiliproteins), enzymes and isotopically labelled compounds. By screening the algal literature for publications on the various uses of microalgae, the amount of papers on the utilization of algae as therapeutics – published especially during the early stages of algae production – gives the impression that this application is one of the most important. Several of these studies have been performed in the Far East, a fact that is not surprising because in this area the cultivation of algae and their utilization in folk medicine has a long tradition.

Although there are still several companies actively searching for pharmaceuticals among microalgae, an immediate breakthrough of any algae-based product cannot be seen. In the following section, selected reports on the therapeutic use of algae are summarized to give an impression of the multiplicity of their application.

14.1.   Therapeutic uses

Many algae either have been, or are, used in folk remedies. The active constituents are generally unknown, although work has been conducted to isolate a few of them in some cases. Several beneficial effects reported for marine algae can be explained by their vitamin and mineral content. Some species, e.g. *Chondria sanguina*, have been used as ingestion remedies, related to the anti-ulcerogenic properties described for carageenans. Other species have been claimed to have hypoglycemic effects; however, the mechanisms involved in lowering blood sugar levels are not understood yet.

Certain marine algae are being used as dressings, ointments, in gynae-cology, in anti-coagulation and anti-hyperlipemic activities, as enzyme activators, sources of vitamins or in the traditional medicine in many countries for the treatment of intestinal helminth or roundworm infections.

Although there are numerous publications on therapeutic properties of marine algae, very little information is available on similar applications of fresh-water microalgae.

Some of the limited reports found in the literature shall now be summarized. It should be stressed that in several cases the beneficial effects were very small and that rigorous scientific controls were seemingly not applied. In addition, it is often difficult to attribute the therapeutic effect clearly to the action of the microalgae used. On the other hand, the scarcity of warranted information does not preclude the possibility that certain microalgal species may posses properties that are of distinct therapeutic value.

Pharmaceutical preparations containing *Spirulina* (whole algae or extracts) as an active ingredient accelerated the cicatrization of wounds (Clement, Rebeller & Zarrouk, 1967a). Treatment was effected with creams, ointments, solutions and suspensions. Other reports showed that *Spirulina* and its enzymatic hydrolyzates promote skin metabolism and prevent keratinization. In Czechoslovakia, the effect of ointments mixed with 20% alcoholic extracts of *Scenedesmus obliquus* was tested on 109 patients suffering from trophic and varicose ulcers, burns, non-healing wounds or eczema (Safar, 1975). According to the author, about 90% of the patients were healed and in 7% an improvement could be observed, whereas out of 112 control individuals, suffering from the same ailments and treated with placebo, only one person improved. The stimulating effect of the algal ointment on granulation and epithel formation was attributed to the chlorophyll, the carotenoids and the B vitamins present in the preparation. Another study describes the production and pharmaceutical application of *Scenedesmus*-containing preparations for the treatment of a variety of skin diseases. Based on the reports collected from physicians and veterinary surgeons, the best results were achieved for the treatment of eczema with children, ulcera cruris, and gynaecological diseases (Rydlo, 1973).

There are several reports from Japan describing different therapeutic effects of *Chlorella* on patients suffering from a long list of diseases. Although the findings show encouraging attributes of the algae, it should be remembered that the investigations were initiated by the *Chlorella*-producing industries so that a certain amount of publicity cannot be excluded completely. One of these reports deals with the administration of *Chlorella*-containing tablets in cases of incurable wounds, describing a tendency to an improvement in granulation of tissues and promotion of epithelium formation (Saito & Oka, 1966).

Quite different aspects were investigated by the Japanese Navy, who tested the effect of *Chlorella* on changes of weight of healthy adults and the morbidity rate of the common cold during a voyage (Kashima & Tanaka, 1966). The application of 2 g of algae per day showed a positive effect on the weight gain of sailors working below deck and reduced the moribidity rate of colds by 30% in the experimental group taking the algal preparation.

One of the beneficial therapeutic properties attributed to *Spirulina* is an appetite-reducing effect, which allegedly helps obese persons in controlling their weight. In order to test the effect of this alga on body weight as well as several other clinical and biochemical parameters, a double-blind cross-over study was carried out on 16 overweight persons (Becker *et al.*, 1986). Patients, who had first taken the algal tablets were given a placebo during the second period and vice versa; each period lasted for four weeks with a 14-day wash-out period between. Fourteen tablets (200 mg algae each) were taken before meals thrice daily. During the administration of the placebo, a statistically insignificant average weight loss of 0.7 kg occurred, whereas during the four weeks of taking the algae the weight dropped significantly by an average of 1.4 kg. However, by comparing the weight loss during the placebo and the verum phase, no significant differences could be established. Under *Spirulina* application, the limit of significance had already been reached within two weeks, but not during the placebo phase. Within the test phase no alarming changes appeared in either clinical, biochemical or subjective findings, and all parameters remained within the normal range.

As mentioned before, *Spirulina* is a valuable source of linolenic acid, which cannot be synthesized by animals or humans. This essential fatty acid has been connected with the stimulation of prostaglandin synthesis. The term prostaglandin summarizes a large number of compounds derived from the 20-carbon polyunsaturated fatty-acid arachidonic acid. In the strictest chemical sense, the term prostaglandin refers to the derivates of the dihomo-γ-linoleic acid, arachidonic acid and eicosapentaenoic acid, which are the precursors of mono-, bis- and trienoic prostaglandins. Prostaglandin E-2 is formed from dietary linoleic acid, which enzymatically is converted to linolenic acid and in turn to dihomo-γ-linolenic acid (8, 11, 14-eicosatrioneic acid). Arachidonic acid, which is actually a metabolic product of linoleic acid, can be converted to prostaglandin E-2. These few examples may demonstrate the significance of a sufficient dietary uptake of these important essential polyunsaturated fatty acids.

Saturated fatty acids as well as other factors can lead to γ-linolenic acid deficiency and suppressed prostaglandin formation. The important functions of prostaglandins result from their effects as local hormones or local regulators of physiological processes, including changes of blood pressure, renal sodium and water excretion, insulin release, gastric acid secretion, etc. Dietary γ-linolenic acid supplementation provided by the uptake of

Spirulina may help to prevent symptoms ascribed to γ-linolenic acid deficiencies.

As shown by many investigators, a variety of non-specific, active immunostimulants can affect the growth of either spontaneous or transplanted tumors in animals or prolong the survival of cancer patients. Immunostimulants exhibiting such effects include various synthetic products, bacteria or bacterial products. In a study, performed in Japan, the antitumor activity of *Chlorella vulgaris* was examined (Tanaka *et al.*, 1984). It was shown that the growth of methyloanthrene-induced fibrosarcomas in a syngeneic or semisyngeneic host was inhibited by injection of a hot-water extract of *Chlorella vulgaris* into the tumor or into subcutaneous tissue near the regional lymph nodes. Both T-cells and macrophages appear to participate in the antitumor effect of the algal extract.

It is well known that several marine algae produce antibiotics. One of the common compounds is acrylic acid, which inhibits the growth of Gram-positive and, to a lesser extent, Gram-negative microorganisms. Phenols, common in macroalgae, are known for their antimicrobial activity, and have also been identified in microalgae; such compounds are, for instance, aplysiatoxins. Certain fatty acids have been attributed to the antibiotic activity of chlorellin from *Chlorella*, and extracts from *Ochromonas*, *Chlamydomonas* and other microalgal species. Also eicosapentaenoic acid, which is synthesized by some algae, shows antibiotic properties.

Following the modern trend, it is not surprising that algal compounds that show inhibition on the cytopathic effects of the HIV-1 virus, which is implicated as a causive agent of AIDS, have been described recently (Gustafson *et al.*, 1989).

As part of a program performed by the National Cancer Institute of the United States with the aim of discovering new antitumor and antiviral agents in natural sources, extracts of various cyanobacteria were tested on human cells for their protective property against HIV-1 infections. A number of extracts were found to be remarkably active against the AIDS virus. Active agents, consisting of sulfolipids with different fatty acid esters (Fig. 14.1), were isolated from *Lyngbya lagerheimii* and *Phormidium tenue* by a combination of gel-permeation and reversed-phase chromatography. They constituted of about 10% by weight of the organic extracts from the cells and were active over a broad concentration range ($1-100 \mu g \, ml^{-1}$). All of the sulfolipids tested had similar levels of activity, suggesting that acyl-chain length and degree of unsaturation in the tested range do not affect their potency. Structurally related acyl glycerols, complex lipids and simple sulfonic-acid derivatives did not protect again HIV-1 infections.

Sulfonic-acid-containing lipids were first described by Benson, Daniel & Wiser (1959). Members of this structural class are commonly referred to as sulfoquinovosyl diacylglycerols that are structural components of chloroplast membranes, commonly found in algae and higher plants. Structurally

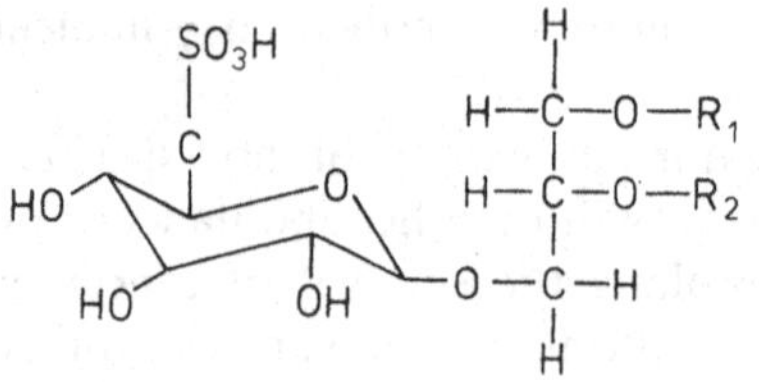

Fatty acid esters

	R_1	R_2
1	18:3	16:0
2	18:2	16:0
3	18:1	16:0
4	16:1	16:0

Fig. 14.1. Chemical structures of generalized depiction of different sulfolipids with R_1 and R_2 showing carbon-chain length.

related acyl glycerols, complex lipids, detergents and simple sulfonic-acid derivates did not show any protective potential. Additional cultured cyanobacterial extracts with inhibitory properties were also found in *Phormidium cebennse, Oscillatoria raciborskii, Scytonema burmanicum, Calothrix elenkinii* and *Anabaena variabilis*. Further studies on these non-nucleoside compounds are already being performed in order to test their feasibility as candidates for comparative clinical testing programs with other known active nucleoside agents.

Serum-free culture of mammalian cell lines has become an effective method in *in vitro* cellular and developmental biology and for the production of monoclonal antibodies and interferon. The cultivation of these cells requires complex culture media containing different growth-promoting factors such as hormones or transferrin etc., originating from animals. Earlier studies have indicated that at least three factors that promote the growth of hybridoma, lymphocyte and tumor cells in serum-free media also could be isolated from the extracts of the thermophilic cyanobacteria *Synechococcus elongatus* and *Spirulina subsalsa* (Shinohara *et al.*, 1986).

More recent studies revealed that some of these growth-promoting substances in the dialyzate from *Synechococcus* are phycobiliproteins: phycocyanin and allophycocyanin. The activity of allophycocyanin was found to be higher than that of phycocyanin. In addition, commercially available phycocyanin from *Spirulina platensis* that may be a mixture of both pigments, also showed growth-promoting activity suggesting that the phycobiliproteins from all kinds of cyanobacteria can be a growth factor for such cell lines. The chromophore of phycobiliproteins is a tetrapyrrol, bound non-covalently to polypeptides, resembling the structures of bile pigments. If these chromophores are the active constituents in this case, there may be the possibility that biliverdin or another phycobiliprotein such as phycoerythrin also has growth-promoting activities.

These preliminary findings suggest that microalgae might be a useful tool for detecting growth-promoting substances required for the *in vitro* cultivation of mammalian cells.

One of the latest attempts to popularize algae is their promotion as a valuable source of carotenoids, which are mostly used as natural food color, as additives for animal feed to enrich the color of their products (egg yolk, poultry and fish meat), or to enhance their health or fertility. Among the over 400 known carotenoids, very few are used commercially for the above-mentioned purposes, for instance β-carotene, lutein, zeaxanthin, astaxanthin, lycopene and bixin. The nutritional and therapeutic relevance of certain carotenoids is their ability to act as provitamin A, i.e. the dietary carotenoid is converted into vitamin A by the consumer.

As the demand for β-carotene especially is higher than the production from natural sources, synthetically produced β-carotene is offered by various companies. β-Carotene is the only carotenoid that has the potential to form two molecules of vitamin A (retinol). All the other about 50 carotenoids that also can serve as precursors for vitamin A cannot form more than one molecule of vitamin A for each provitamin A carotenoid in mammals. In humans, the conversion of β-carotene into vitamin A occurs in the small intestine where both β-carotene and vitamin A are absorbed and found in plasma, tissue, and organs of the body.

For practical purposes it is assumed that β-carotene comprises all of the provitamin A carotenoids. The average daily β-carotene intake is about 1.5 mg, equivalent to 250 retinol equivalents or 2500 IU of vitamin A. It is further assumed that the overall utilization of β-carotene as a source of vitamin A is one-sixth that of retinol.

In view of the nutritional importance of β-carotene and the growing utilization of certain microalgae as a source of this pigment, a few additional explanations shall be included here. The unicellular halotolerant alga *Dunaliella bardawil* was previously shown to contain high concentrations of β-carotene, composed of the all-*trans* and the 9-*cis* isomers. The physico-chemical properties of the 9-*cis* form are different in several aspects from those of the all-*trans* isomer. Especially noteworthy is the higher solubility of the 9-*cis* form in hydrophobic solvents and its lack of crystal formation. In other words, the all-*trans* isomer is almost insoluble in oil, whereas 9-*cis*-β-carotene is much more soluble in fats, thus leading to higher absorption and storage in animal tissues.

Natural β-carotene, as found in many algae, especially in strains of *Dunaliella*, as well as in many fruits and vegetables, contains about 50% all-*trans* β-carotene, with the rest composed mostly of 9-*cis*-β-carotene (Fig. 12.1). Both forms are accumulated in the liver in a ratio similar to that present in the dietary β-carotene. Although some of the studies performed to support a connection between β-carotene intake and reduced incidences of human cancer (see below), used natural β-carotene from fruits, most of the studies used synthetic all-*trans* β-carotene.

There is epidemiological and biological evidence that carotenoids may serve as chemopreventive agents against certain types of cancer, and preliminary clinical evaluations indicate that β-carotene (as provitamin A) and other carotenoids (canthaxanthin, phytoene) prevent photosensitivity when given to patients suffering from light-sensitive skin diseases. For instance, β-carotene is available as a prescription drug (daily dosages of 30–180 mg) for the treatment of patients who suffer from the inherited disease of abnormal photosensitivity reactions in the skin (erythropoietic proto-porphyra). Under this treatment no adverse effects (i.e. hypervitaminosis) have been noted so far.

Animal experiments suggest that carotenoid pigments, irrespective of their vitamin A activity, can prevent or slow down the growth of skin tumors induced by UV-light. Retrospective or prospective epidemiological studies, concerned with the effect of diet, particularly the ingestion of carotenoid-containing vegetables, have shown that there is an inverse relationship between the ingestion of carotenoids and the incidence of cancer (Mathews-Roth, 1985).

Results published in the literature are often contradictory. There are reports suggesting that carrots, which contain primarily α and β-carotene, both precursors of vitamin A, are not associated with a protective action, whereas tomatoes, which contain nutritionally inactive lycopene, or dried fruits were. Some authors come to the conclusion that the intake of preformed vitamin A by persons with an adequate vitamin status does not seem to be associated with any protection against cancer; carotenoids as a whole, however, tend to be more promising as possible protective agents, whereas other authors do not distinguish between carotenoids as a whole and specific pigments.

It has been demonstrated that photosensitized reactions in the skin involve the formation of singlet oxygen. The main protective and therapeutic utility of the carotenoids seems to be their ability to quench singlet oxygen, providing protection against neoplasm. Although humans accumulate carotenoids and retinol in serum and tissue, only β-carotene, but not retinol, can serve as an antioxidant as well as a potent quencher of the highly reactive molecular singlet oxygen.

It has to be kept in mind that chemopreventive agents are commonly ingested over a long period of time to obtain maximal protection and as a kind of insurance that the risk of contracting a specific malady will be reduced. However, an agent useful in preventing one sickness may actually enhance the risk of contracting another, for instance carotenodermia in this context.

In our society, vitamin A deficiency is extremely rare except as a result of certain genetic or chronic diseases. Therefore, the question remains whether amounts of vitamin A above that needed for nutritional adequacy have any effect.

At present, several studies with humans on the protective properties of carotenoids are being performed but it will take quite a few more years before it is definitely established whether or not β-carotene and other carotenoids have a significant anticancer effect in man.

Besides carotenoids, the water-soluble pigment phycocyanin, which under certain growth conditions may constitute more than 20% of *Spirulina*, has also been attributed with cancer-preventing properties. It was reported that in a study with tumor-induced mice the survival rate in animals fed with a phycocyanin extract was significantly higher than that of the controls (Richmond & Becker, 1986). In further studies with mice it was observed that lymphocyte activity in the treatment group was higher than that in the control group. According to the authors of the study, phycocyanin may generally stimulate the immune system, providing protection from a variety of diseases.

Summarizing, it may be stated that it is too early to attribute clearly a therapeutic chemopreventive anticancer effect to microalgae as a source of carotenoids or phycocyanin. We have to await the results of on-going clinical studies and others that may also commence in the near future before it can be decided whether the above pigments will play an important pharmaceutical role for use in large populations.

Many deaths are caused by heart diseases that are associated with high plasma cholesterol levels and hypertension. Some marine algae such as *Cystoseira barbata*, *Fucus gardnerii* and *Phyllophora nervosa* have been shown to lower plasma cholesterol levels, and in some cases the active compounds have been identified. Unsaponifiable sterols and unsaturated fatty acids, for instance, showed hypocholesterolemic activities. The effects of eating marine green algae (*Monostroma nitidum*, *Ulva pertusa*, *Enteromorpha compressa* and *E. intestinalis*) on cholesterol metabolism have been studied in rats. All algae were found to lower significantly the cholesterol levels, most probably because of betaines present in these algae.

During the performance of nutritional studies with *Scenedesmus obliquus* on rats it was observed that animals fed algae-containing diets showed lower cholesterol levels than the controls. Rolle & Pabst (1980), who tested the cholesterol-reducing activity of drum-dried *Scenedesmus*, found that in animals fed for six to eight weeks with a standard diet enriched with 3% cholesterol, the average concentration of blood plasma cholesterol increased from 2.0 to 3.6 mmol l^{-1}. However, in animals receiving the cholesterol in a diet enriched with 20% algal powder, the final plasma cholesterol level only slightly increased to 2.4–2.8 mmol l^{-1}, whereas in animals fed solely algae, the cholesterol concentration increased by 0.2 mmol l^{-1} only. The level of plasma triglycerides in animals fed on algae with and without the addition of cholesterol was lower than that of the controls fed algal-free diets; the algae-enriched diet prevented an excessive deposition of cholesterol in the liver. The same authors also studied the effect of hydrophilic and lipophilic algal extracts and the remaining algal residues for their cholesterol-lowering properties. The different fractions were obtained by hot-water treatment and chloroform/methanol extraction. The algae extracted with water lowered the plasma cholesterol levels; the content of plasma triglycerides in animals receiving the different fractions was reduced by 35–55% in nearly all groups. The cholesterol content in the liver was reduced up to 50% by untreated algal powder as well as by algal material extracted with water or organic solvents; actually, the remaining algal residue after extraction showed the best effects (Table 14.1). From these observations it seems that the crude fibers of the algae represent the effective component that lowered the cholesterol level, an observation that is also known from marine algae.

Similar experiments with *Scenedesmus* and *Spirulina* were reported from India (Anusuya Devi & Venkataraman, 1983a). These authors also employed rats to test a possible hypocholesterolemic effect of different diets, containing algae at 10% and 15% protein level and casein (10% protein) as control. Serum and liver cholesterol levels after a feeding period of six weeks are listed in Table 14.2. The serum cholesterol level was highest in rats fed casein; incorporation of *Scenedesmus* at both protein levels significantly lowered the serum and liver cholesterol levels. The cholesterol-lowering property of sun-dried *Spirulina* was not as pronounced as that of drum-dried *Scenedesmus*. Because the constituents of both algae have not been evaluated separately, it is difficult to explain the lesser effect of *Spirulina*. However, as previously mentioned, it is quite possible that the somewhat higher carbohydrate and cellulose concentrations in *Scenedesmus* are the main causes of the cholesterol-lowering property observed with this alga in the animal-feeding studies.

More recently, the effect of *Spirulina* on serum lipids was studied in 30 healthy male volunteers who had mild hyperlipidemia or mild hypertension (Nakaya, Homma & Goto, 1988). The test individuals were divided into two groups. Members of the first group were given 4.2 g of *Spirulina* per day for eight weeks; members of the second one were given the same amount of the alga for the first four weeks, and for the next four weeks they were observed without *Spirulina*. Mean total serum cholesterol of the first group was reduced significantly from 24.3 ± 3.6 mg ml^{-1} at the beginning of the observation period to 23.3 ± 3.6 mg ml^{-1} after four weeks. This reduction was maintained over the whole experimental period of eight weeks. A similar reduction rate was seen in the second group; when administration was discontinued, the cholesterol levels returned to the initial values. The cholesterol-lowering potential of the alga was still more pronounced with persons having serum cholesterol levels over 22.0 mg ml^{-1}. No adverse

Table 14.1. *Liver weight and cholesterol concentration of rats fed cholesterol-free (−Chol) and cholesterol-containing (3%Chol) diets enriched with alga (drum-dried* Scenedesmus*) and algal extracts*

Diet	Liver weight (g dry matter)		Liver cholesterol (% of dry matter)						Esterified cholesterol (% of dry matter)	
			Free		Total					
	−Chol	3%Chol	−Chol	3%Chol	−Chol	3%Chol	−Chol	3%Chol	−Chol	3%Chol
C	6.1	7.3	0.54	0.57	0.77	7.13			0.23	6.55
A	5.7	6.4	0.60	0.69	0.75	3.68			0.15	2.99
SE	5.2	7.6	0.74	0.46	0.90	3.49			0.16	3.03
ASR	6.6	8.5	0.56	0.40	0.77	5.37			0.21	4.97
WE	6.4	8.0	0.53	0.50	0.78	3.55			0.25	3.05
AWR	6.6	8.6	0.55	0.66	0.77	7.10			0.22	6.14

Notes:
C, Control diet.
A, Algal diet.
SE, Solvent algal extract (chloroform/methanol 2:1).
ASR, Algal residue after solvent extraction.
WE, Hot-water algal extract.
AWR, Algal residue after hot-water extract.

Table 14.2. *Hypocholesterolemic effect of diets containing drum-dried* Scenedesmus *and sun-dried* Spirulina

Diet	Protein (%)	Serum cholesterol (mg per 100 ml serum)		Total liver cholesterol (mg per 100 g fresh liver)
		Total	Free	
Casein	10	251.9	75.0	47.6
Scenedesmus	10	121.2	50.7	39.7
Scenedesmus	15	94.2	32.7	34.3
Spirulina	10	233.5	65.6	43.8
Spirulina	15	220.0	53.0	36.4
Spirulina + 0.3% met	10	229.0	56.0	40.6

effects were observed during this study. Although these results are still premature, they point to a direction where a therapeutic application of algae might be of real use.

14.2.   Heavy-metal removal

For several years it has been repeatedly suggested by many authors that heavy metals from polluted aqueous systems may be removed by phytoplanktonic algae. It is concluded that this method, including the separation of the metal-saturated algae from the medium, is an economic method for removing heavy metals from waste waters, resulting in high-quality reusable effluent water and valuable algal biomass, which could be used for different purposes (production of biogas, fertilizer, fodder, etc.). It is well established that several marine and fresh-water algae are able to take up various heavy metals selectively from aqueous media and to accumulate these metals within their cells. Although impressive accumulation factors in the order of several thousands have been reported for certain microalgae, which look very promising for this purpose, it has to be asked how far these optimistic expectations are practicable. The following reflections and a short calculation based on realistic figures and inputs should elucidate the restrictions of this proposal.

There is no doubt that the heavy-metal contents in urban waste waters, although they do not reach the proportion found in industrial effluents, and certainly not those of metal processing industries, are of public concern because they cause several problems, particularly in areas with a dense population. Some of the constraints caused by heavy metals are, for instance, their unfavorable effects on biological sewage treatment plants by inhibition of nitrification and interference in biological oxidation.

In general, three different types of pond in waste-water treatment can be distinguished: anaerobic, facultative and aerobic. Anaerobic ponds are several meters deep, they are free of dissolved oxygen and have high BOD removal rates. Facultative ponds are the most common form of oxidation ponds; they have aerobic conditions on the surface because of photosynthetic oxygen production by microalgae and anaerobic conditions in the bottom layers. Aerobic ponds, also called high-rate ponds, are shallow and completely oxygenated ponds that are best suited for algal growth.

The following computations are based on field experiments performed to study the effect of different sewage-treatment processes on heavy-metal absorption by fresh-water algae under natural conditions (Becker, 1983). For this the algae were separated from the medium by a semi-permeable membrane, which allows diffusion of nutrients and heavy-metal ions but retains the algal cells. In connection with the problem discussed here it seems to be irrelevant, if the concentration of heavy metals by algae is based on active or passive processes, i.e. whether it is adsorption to the cell surface, passive penetration into the cell or energy-requiring accumulation.

The efficiency of the concept of using algae for heavy-metal removal will be determined principally by the following five parameters: 1) growth rate of the algae; 2) metal concentration factor of the algae; 3) concentration of heavy metals in the medium; 4) desired percentage of metal removal from the medium; and 5) metal recovery in relation to capital and operating costs.

1) As detailed elsewhere, the reported data of algal growth and yields vary to a certain extent because of various intrinsic and extrinsic factors. Assuming optimum conditions, the upper yield limit can be expected in the range of 20–40 g dry weight $m^{-2} d^{-1}$. It is this figure that has to be considered in estimating the feasibility of any large-scale outdoor algal cultivation venture. In addition, the number of algal species that are predominant in sewage is limited to a few only, such as *Chlorella* spp., *Scenedesmus* spp., *Micractinium* spp. and *Oscillatoria* spp. For the present calculation, an average yield of 30 g $m^{-2} d^{-1}$ is assumed for variable pond depths, growth retardation due to nutrient limitation, infections, fluctuations of algae population, etc.

2) A quick glance through the relevant literature illustrates that algal heavy-metal accumulation factors vary by several orders of magnitude. However, in order to simplify this actually very complex aspect, it is assumed that an accumulation factor of 5000 can be maintained, keeping in mind that this factor depends on algal density, type and concentration of heavy metals and their interference with other ions, and environmental conditions.

3) Although the concentrations of heavy metals in waste water fluctuate considerably, in the present case a total concentration of 1 mg l^{-1} is assumed. An increase of this figure will increase the treatment time proportionally, as long as the metal concentration does not reach toxic levels.

4) The concentration of heavy metals in the effluent of sewage-treatment plants is not stipulated in several countries so that no definite figures shall be given here. On the other hand, the envisaged final concentration can be extrapolated from the graphs in Figs 14.2 and 14.3.

   The minimum exposure time of the algae with the metal ions to attain intracellular saturation is determined by the velocity of the accumulation through the algal cell, which is a relatively rapid process of about 24 h. Any exposure time longer than this will not improve the accumulation capacity of the algae. Hence a maximum exposure time of one day is assumed in this context.

5) Finally, the removal of heavy metals from waste water by means of algae has to be seen also in the frame of economic aspects.

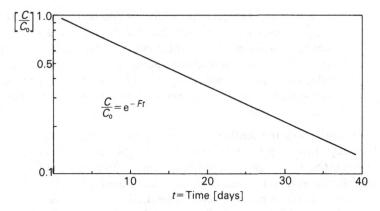

Fig. 14.2. Heavy-metal removing efficiency in a static system. Time course of reduction of the initial metal concentration C/C_0.

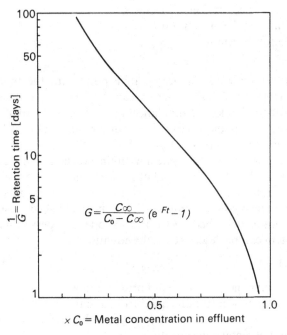

Fig. 14.3. Heavy-metal removing efficiency of algae in a dynamic system. Dependency of retention time (part of total pond volume renewed daily $= G$) on reduction of initial metal concentration.

It is an absolute prerequisite that biological treatment is feasible only in cases where no additional investments are necessary to convert traditional treatment ponds into high-rate ponds. However, this is the most crucial point in this concept because an effective metal removal by algae requires long detention times and consequently large ponds, which are not common in conventional treatment plants.

In evaluating the feasibility of heavy-metal removal by algae, two basically different systems have to be distinguished: a) a static, and b) a dynamic (continuous/semi-continuous) one. A static model implies a pond with the aforesaid accumulation factors and growth rates to which no fresh waste is added for a given period of time. In this case the efficiency, i.e. the ratio of initial metal concentration to remaining concentration after partial removal, can be expressed as:

$$C_t = C_o \times e^{-Ft},$$

where C_t = metal concentration after t days
C_o = initial metal concentration
F = dilution constant
t = time.

If we apply this equation to a representative pond with the following parameters: volume $= 10\,000$ m³, surface area $= 3600$ m², daily algae production (dry matter) $= 100$ kg, accumulation factor $= 5000$, $F = 0.05129$, a detention time of about 14 days is calculated for a reduction of the heavy-metal concentration by 50% (Fig. 14.2).

The dynamic model describes a system wherein a certain amount of the pond value is renewed daily by removal of treated, and addition of fresh, waste. In this model the relation between the amount of water added daily to the percentage of heavy-metal reduction has to be established. Assuming the same conditions as described for the static model, the performance of the dynamic system can be described by the equation:

$$C_{n+1} = (1 - G) \times C_n \times e^{-Ft} + G \times C_o \times e^{-Ft},$$

where C_n = metal concentration equilibrium at day n
C_{n+1} = metal concentration equilibrium at day $n+1$
G = part of total volume renewed every day (retention time)
C_o = initial metal concentration.

By setting C_n and $C_{n+1} = C_\infty$ as the limit of the function:

$$C_\infty = \frac{G \times C_o \times e^{-Ft}}{1 - (1 - G) \times e^{-Ft}}$$

and $$G = \frac{C_\infty}{C_o - C_\infty} \times (e^{Ft} - 1)$$

It can be seen that in the dynamic (semi-continuous) model a constant 50% reduction of the metal concentration can be achieved by a retention time of 19 days, corresponding to a daily removal of $\frac{1}{19}$ of the pond volume (Fig. 14.3).

Although according to this calculation the static model seems to be more efficient per time unit, it is not to practical for various reasons. The amount of nutrients and biodegradable compounds in the sewage is not normally sufficient to maintain continuous optimal algal growth without the addition of nutrients (mainly nitrate and phosphate, which are 5–10 times lower than in conventional algae cultivation media).

To maintain continuous optimum growth rates a certain amount of algal biomass has to be removed every day in order to minimize growth inhibition due to light limitation caused by self-shading effects. In a conventional treatment plant, fresh waste water accumulates every day so that in the static model sufficient additional ponds have to be provided to hold this amount of fresh waste during the treatment phase of the first pond. The management of such series of ponds and the daily harvests of algae are technical problems and an important cost factor. Under these circumstances the dynamic system seems to be more feasible in spite of the theoretical lower metal removal efficiency because it offers the possibility of operating with one pond only. The necessary daily removal of waste in this system can be combined with the separation of the algal biomass. This harvesting step is another crucial technical point and a costly part of the concept of metal removal by bioabsorption. Among the various harvesting systems available, only flocculation seems to be the acceptable method for this purpose. If alum is used as the flocculant, a ratio of flocculant needed to initial algae biomass concentration of 0.5 has to be assumed to achieve 80% algae removal.

By analyzing critically the described systems it can be concluded that at present none of the available methods is suitable for a practical and economical application. There are too many unsolved problems, i.e. required size of the treatment plant, necessary time for effective metal removal, problems in controlling and maintaining optimum algal growth and separation of the algae. In addition, it should also be kept in mind that any impairment of the assumed optimal conditions will further reduce the efficiency of the metal removal envisaged. The short detention times customary in conventional treatment plant are too short by far for bioabsorption.

Heavy-metal removal from polluted water by means of algae might be suitable in special, limited cases where small volumes of water have to be treated and where the detention times are of secondary importance.

14.3.   Immobilized algae

In the past 20 years, the use of immobilized enzymes or cell components for the production of a series of metabolites has become a branch of biotechnology of rapidly growing importance. Although in the initial stage most of the research work on immobilization dealt with systems designed for the release of products, synthesized by enzymes or multi-enzyme complexes, a more recent development focuses on the immobilization of complete cells or cell agglomerates. To a certain extent these systems resemble natural environmental conditions as many microorganisms grow in a biotiope where they are also immobilized by encapsulation in slimes or as a partner of symbiotic systems. Although the pioneering work with immobilized cells mostly employed heterotrophic organisms, a number of scientific reports today deal with studies on plant cells, algae, cyanobacteria and photosynthetic bacteria. These phototrophic microbes offer several prospects for use in immobilization techniques because they can use sunlight as their sole, or major, energy source to make products from the substrates of photosynthesis.

It is impossible for open algae cultivation systems to be used for the production of bioproducts such as vitamins, polysaccharides or enzymes, because such a plant will neither be uni-algal nor will it be possible to extract the bioproducts from the medium; they will be used up or catabolized by other microorganisms present in the culture.

A more promising approach for recovering excreted algal products would be the autotrophic or phototrophic cultivation of axenic algal cultures in enclosed bioreactors such as transparent plastic tubes or illuminated fermentors. Another possibility is the culture of algae in closed systems, but immobilized on a surface or in a gel. This method has the advantage that a good ratio of algal biomass to medium volume can be achieved, thus reducing the total volume of medium from which the desired bioproduct has to be extracted.

This possibility led to a more recent approach in applied algology, i.e. the preparation and utilization of microalgae in an immobilized form (Robinson, Mak & Trevan, 1986).

It has been known for many years that algae are involved in the purification processes occurring in percolating filters of waste-water treatment plants, where they can be found as thin films in the form of encrustrations or sheets in the upper zone of filter beds. These intrinsic tendencies of algae to adhere to surfaces or to flocculate has been taken up for the development of various techniques of preparing immobilized algal cells.

Algae excrete a wide range of secondary metabolites including organic acids, amino acids, peptides and polysaccharides. Algal cells thus retain the essential virtue of microbial cells to excrete many of their metabolites, while exhibiting many characteristics of plant cells. Furthermore, immobilization of algae is also of interest in that it provides a model for the immobilization of higher-plant cells and enables the effects of immobilization on cell physiology *per se* to be examined.

During recent years, several techniques have been developed for immobilizing algae and keeping them viable for long periods of time, i.e. adsorption, entrapment in polymers or gels, covalent coupling and cross linking to insoluble matrices. This can be done by either active or passive entrapment. In the first method, precultured algae are fixed within the polymeric matrix, whereas the latter technique simply depends on algal growth to invade the matrix.

Various optical- and electron-microscopy studies have been carried out on the growth and morphology of immobilized algal cells and most have confirmed that the entrapment alters the morphology or algal cell function very little, but allows better separation of biocatalyst and product.

The materials being considered for the immobilization of algae cells have to have as many as possible of the following properties.

> High biomass loading capacity.
> High transparency.
> Absence of toxicity to the algal cells.
> Suitable for simple immobilization procedures.
> Optimum diffusion of nutrients and products.
> Resistance to abrasion.
> Maximum surface-to-area volume.
> Mechanical stability.
> Sterilizable.
> Suitable for easy removal of media from matrix.

Various natural and synthetic polymers have been tested for the entrapment of algae. The fundamental difference between the materials used for the immobilization of algae and the materials employed for the immobilization of bacteria, yeasts and fungi is the necessity that, for algae, materials have to be chosen that are transparent or translucent to permit deep penetration of light into the bioreactor in order to maintain the photosynthetic processes of the algae.

The natural compounds frequently used for this purpose are alginate, agar, agarose, carrageenan or serum albumin glutaraldehyde, whereas polyurethane, polyvinyl foams, acrylamide, ceramic or glass beads are the current synthetic materials. Most of the natural polymers ensure high algae viability, whereas acrylamide and serum albumin have been found unsuitable for certain applications because of toxic effects on the algae.

Different methods for the immobilization of algae have been described; the most popular technique is entrapment in porous gels and foams made from polymers. The most commonly used synthetic polymer is polyurethane, although considerable amounts of algae are destroyed while preparing the matrix with the toxic compound isocyanate. This problem can be partially overcome by intensive rinsing with the culture medium. Commercially available polyurethane blocks or glass beads have also been used for passive immobilization techniques by simply immersing the material in growing algal cultures.

For the immobilization of algae in polyurethane foams (polyester or polyvinyl type), small cubes ($0.5 \times 0.5 \times 0.5$ cm) of foam are cut and washed three to five times in distilled water for a few days by carefully removing the bubbles from the foam every time. About one cube per milliliter of algal growth medium is placed into the flask containing the culture medium, sterilized by autoclaving and then inoculated with the algae culture. Depending on the algal growth rate, the cubes can be removed after about one week, rinsed and aseptically resuspended with fresh medium and then used for further studies in suitable reactors.

Cell envelope components are probably solely responsible for cell hydrophobicity, which plays the major role in adhesion of benthic cyanobacteria on solid surfaces that have little or no surface charges.

Alternatively, immobilization can also be performed by mixing equal amounts (w/w) of concentrated algal suspension with urethane prepolymer (tolylene diisocyanate hydrophilic prepolymer) for 1 min in an ice bath and then leaving the mixture for 15 min at room temperature for polymerization without further mixing. After that the foam is cut into small cubes of about 0.5 cm edge length and rinsed several times with culture medium.

For immobilization in agarose, algae cells, collected from exponentially growing cultures by centrifugation, are suspended in fresh culture medium, mixed with 5% commercially available agarose at about 40 °C and then poured, with stirring, into vegetable oil heated to the same temperature. The drops so formed are solidified by cooling the mixture to 5 °C. The beads have to be filtered off and washed free of oil.

For immobilization of microalgae in carageenan, one part of algae suspension is mixed with three parts of carrageenan solutions at 38 °C to obtain a final concentration of 2.5% carrageenan. The mixture is dropped into a gently stirred 2% KCl solution at 20 °C, where the beads are soaked for 30 min to increase their stability. Agar is another natural compound that has occasionally been used for algal immobilization, performed as follows. Equal amounts of Tris-HCl buffer (0.1 M, pH 8.0, 50 °C) containing 4% agar, and algae suspended in Tris-HCl buffer, are mixed and immediately cooled to room temperature. The solidified gel is cut into small pieces.

Alginate is the most common entrapment material used for the immobilization of algae cells because it stands out as the most promising and versatile method yet. The immobilization can be carried out in a single-step process under very mild conditions and is therefore compatible with most living cells. Alginates constitute a family of unbranched binary copolymers of 1–4 linked β-D-mannuronic acid (M) and α-L-guluronic acid (G) of widely varying composition and sequence. In practice it is difficult to select suitable alginates, because most of the producers normally do not specify the chemical composition or algal source of the products. Commercial alginates are produced mainly from *Laminaria hyperborea, L. digitata, L. japonica, Macrocystis pyrifera* and *Ascophyllum nodosum* (Smidsrød & Skjåk-Brœk, 1990). For the production of the alginate solution, a 2–5% (w/ v) aqueous solution of Na-alginate is prepared by dispersing the polymer in distilled water or buffer under extended (at least 6 h) stirring at room temperature. For long-term experiments it is mandatory to work under sterile conditions. Alginate solution can be sterilized by autoclaving (pH 7– 8) or sterile filtration. It should be observed that extended heating may cause some depolymerization of the alginate chain so that filtration (pore size 0.22 μm) is the most preferable method. Alginate charges of inferior quality should be filtered through a series of filters with decreasing pore size $(1.2 \rightarrow 0.8 \rightarrow 0.45 \rightarrow 0.22 \ \mu$m) in order to obtain highly transparent beads. The alginate solution thus prepared is mixed with half the volume of sterile concentrated algal suspension in a buffer solution, avoiding divalent cations. To obtain gel beads the readily mixed alginate/algae suspension is dripped at room temperature through a syringe with cannula diameter between 0.2 and 1.0 mm or a small funnel into a solution containing 20–100 mM $CaCl_2$, in which they should be kept for 10–30 min for hardening at 4 °C. The droplets form gel spheres instantaneously, entrapping the cells in a three-dimensional lattice of ionically cross-linked alginate via Ca^{2+}.

The beads are removed, blot dried, rinsed in buffer several times and given into the respective reactor together with fresh medium; 500 ml of cell suspension and alginic acid solution produce about 250 ml of beads.

The physical characteristics of alginate beads appear to be dependent on a variety of interactive factors. Increased pH and phosphate concentrations show a general trend of bead disruption. Among various cations tested, Ca^{2+} formed the largest beads. The diameter is not markedly affected by the cation concentration or by the concentration of algae cells in the bead, but generally increases with alginate concentration. Gel rigidity is proportional to the square of the alginate concentration when the gel is completely converted to the calcium form. The concentration of phosphate in the medium plays a significant role in the stability of the beads. At pH values above 5.5, alginate beads dissolve within a few days. The choices of conditions that will allow bead stability need to involve considerations of cation type, phosphate concentration and pH value.

Cell leakage is another problem encountered in long-time cultivation of immobilized algae and seems to be caused by chelation of alginate, cell growth and an interaction between both. Therefore, when establishing an immobilized cell system, the priorities must be to inhibit growth and increase the volume of the beads able to support viable cells by reducing the effects of diffusional resistance.

Ions with a high affinity to Ca^{2+} such as phosphate or citrate will sequester the cross-linking Ca^{2+} ion thus destabilizing the gel; this also may occur in the presence of the anti-gelling cations Na^+ and Mg^{2+}. To overcome these problems it is advisable to keep the beads in a medium containing a few millimoles of free Ca^{2+} ions and to keep the $Na^+:Ca^{2+}$ ratio low. Alginate beads can be stabilized by replacing Ca^{2+} by other ions such as Pb^{2+}, Cu^{2+} or Cd^{2+}. However, the use of these toxic compounds is limited to very few applications only.

It seems likely that several immobilized algal systems suffer from rather poor cell retention, which for commercial applications would be quite obstructive because it would necessitate an extra downstream processing step, thus reducing the advantage of the immobilized over the free cell system.

The mechanical resistance of Ca^{2+}-alginate beads varies with the composition of the alginate; beads from alginate with more than 70% α-L-guluronic acid show high mechanical strength. The turbidity of the gel depends on the gel strength; transparent gels are achieved by using alginate with a α-L-guluronic content greater than 60%.

For the production of bioproducts, the diffusion characteristics of the gel beads are important parameters. Diffusion of small molecules is not normally limited by the alginate gel, whereas larger molecules may be retained because of diffusion barriers within the immobilization matrix. The largest pore sizes are found in beads made from high-G alginates.

In conclusion it can be stated that alginate with a high content of guluronic acid is preferable in bioreactors because of its high stability, high porosity, high tolerance to salts, and high transparency, which makes it suitable for the immobilization of photosynthesizing cells such as algae.

In general, two types of algal entrapment can be differentiated: 1) thin gels, less than 1 mm thick; and 2) round matrix droplets or beads with diameters up to 5 mm. To host these immobilized cells, several types of bioreactor have been designed, depending on the purpose the algae are employed for. The most common reactor types for immobilized algae are comparable to the units used in bacteria and yeast fermentation technology.

1) Stirred tank reactors, which ensure effective utilization of light. This is the most widely used reactor in fermentation technology, showing good mass transfer and mixing characteristics. It is equipped with a marine impeller and rounded bottom which improves mixing and lift at low stirring speeds. Media perfusion and withdrawal poses some problems, which partly can be overcome by the use of spin filters.

2) Packed-bed reactors, which often have the disadvantages of poor light penetration, biocatalyst mixing and gas flow because of too-dense packing of the beads and difficulties in removal of the

biomass from the bed. This system shows no particle to particle abrasion and facilitates continuous removal of media.

3) Fluidized-bed reactors, originally designed for waste-water treatment. They tend to result in heterogenous systems but normally show good mixing characteristics without clogging or gas-bubble entrapment.

4) Air-lift reactors, which ensure an almost homogenous distribution of the matrix beads and which are suitable for many different applications.

A drawback of the last two designs is the fact that they do not provide the high matrix (algae) density of plate or packed-bed reactors, which is desired for commercial large-scale application.

The major fields envisaged for the utilization of immobilized algae are as follows:

a) Accumulation and removal of waste products in aqueous systems.
b) Biosynthesis and biotransformation of different natural products such as polysaccharides, enzymes, etc.
c) Production of ammonia.
d) Production of photosynthetic oxygen in combined bacteria–algae systems.
e) Production of hydrogen.

It is well known that several algae species develop impressive capabilities of accumulating certain compounds, for instance heavy metals but also nitrogen (in the form of ammonia) and phosphorus (as orthophosphate), from the environment. It can be shown that species of *Scenedesmus*, immobilized in carrageenan, are able to remove within a few hours high percentages of phosphate and ammonia from typical urban secondary effluents at a similar rate to free-living cells, indicating a possible application of such systems in the tertiary treatment of waste waters. As already mentioned, calcium alginate cannot be used in such a process for the removal of phosphate because this matrix would lose its polymeric structure, so in this context carrageenan constitutes a definite advantage over alginates.

A further improvement of this system and a possibility of reducing the costs could be achieved by using hyperconcentrated algae in order to get a higher cell load in the beads and by using less costly, less purified carrageenan.

Similar results were obtained on the removal of heavy metals in industrial effluents, where the immobilizing matrix seems to protect the algae to a certain degree against toxic effects of the metal ions.

Most of the studies on immobilized algae deal with their use for biosynthesis and transformation of valuable biological compounds such as enzymes, polysaccharides, NADPH, amino acids and hydrocarbons.

A number of microorganisms, showing amino acid oxidase (AAO) activity, have been tested as potential agents for converting amino acids into α-keto acids by oxidative deamination:

$$R-CH(NH_2)COOH + H_2O + O_2 \longrightarrow R-CO-COOH + NH_3 + H_2O_2.$$

The same conversion can also be carried out by algae, either directly by species containing the same enzyme, or indirectly in combination with bacteria, to which the algae will supply the required oxygen.

Glycolic acid production is a characteristic of all plants that fix CO_2 via the Calvin cycle. Studies on free-living algal cells have demonstrated that the excretion of this compound occurs widely among photosynthetic organisms. In *Chlorella* cells immobilized in calcium alginate gel, glycolate production could be maintained over a period of six months. Other research groups have demonstrated the photoproduction of NADP by *Nostoc muscorum*, immobilized in polyurethane foam, the continuous production of amino acids over 10 days by different mutants of cyanobacteria, encapsulated in alginate, the long-term release of sulfated polysaccharides from polyurethane-entrapped *Porphyridium cruentum* or the production of hydrocarbons by *Botryococcus braunii*.

Dunaliella sp. is a marine microalga well-known for its potential to produce glycerol and β-carotene. Reported results suggest significant release of glycerol by this alga depending on parameters such as temperature, light intensity and salinity. Studies on *Dunaliella tertiolecta*, immobilized in Ca-alginate beads, showed to produce significant amounts of glycerol (5 g l^{-1}), even in hypersaline media (up to 4 M NaCl) over a period of several months.

Various other biotransformation tests have been reported using, for instance, amino acids for the production of α-keto acids or radioactive precursors for incorporation into various labeled compounds. In addition to the metabolites mentioned before, the production of other compounds such as pigments, vitamins, growth stimulants and antibiotics are projected as potential products from immobilized algae. However, it seems that all the processes described are still far away from practical and economic commercial utilization because the methods used for these processes are still too costly.

Another application of immobilized algae is the production of ammonia. Its production by N_2-fixing cyanobacteria, immobilized in alginate matrix, was described by using either inactivation of glutamine synthetase by L-methionine-D, L-sulfoximine (MSX) or glutamine synthetase deficient strains, showing a yield of 40% of fixed nitrogen excreted as ammonia.

Detailed comparative investigations on the release of ammonia by the cyanobacterium *Anabaena azolla*, the symbiont of the water fern *Azolla*, have been performed by Shi & Hall (1988). In these studies, the cyanobacterium was immobilized in polyvinyl, polyurethane or alginate, with the aim of evaluating the potential of this system for the production of nitrogen fertilizers. It was observed that the immobilized cells (without MSX-supply) maintained their ability to excrete ammonia even after long-term culture, whereas in the isolated free-living cells this property ceased very quickly. After the addition of MSX, high yields of ammonia could be obtained from immobilized cells. However, studies in different types of reactors have shown that the use of glutamine-synthetase inhibitors limits stable long-term ammonia production because of difficulties in regulating nitrogenase activity and product yield. This problem might be overcome by the use of enzyme-deficient strains. Although the initial results of ammonia production are encouraging, there is still a long way to go until reactors with algal-loaden foam pieces can be found standing next to the fields to provide the required nitrogen fertilizer.

A more promising approach for the utilization of immobilized algae seems to be their employment as a source of photosynthetically produced oxygen in combination with other (also immobilized) heterotrophic, oxygen-requiring microorganisms. Production of oxygen for 25 days was reported for *Chlorella vulgaris* and *Scenedesmus obliquus* after immobilization in urethane prepolymer, and a period of over six months for *Chlorella emersonii* entrapped in alginate gel. Oxygen production was also described for the immobilized red alga *Porphyridium cruentum*.

Stimulated by the increasing costs of energy and the ongoing search for new, unconventional and renewable sources of fuel, different systems for microbial production of hydrogen have been tested, including the use of immobilized algae. Although, for instance, immobilized photosynthetic bacteria require organic carbon sources or inorganic sulfides as electron donors, cyanobacteria are able to use water for the photolytic process to release hydrogen with light as the energy source.

Several investigations on the biotransformation of organic substrates were carried out with immobilized *Botryococcus*, an alga known to produce considerable amounts of hydrocarbons and for some time considered as an renewable source of fuel. With the addition of oleic acid as a substrate to the medium, hydrocarbon production in the range of about $100 \, \mu g \, l^{-1}$ culture could be obtained. Photosystem I of isolated chloroplasts can couple electron transport from ascorbate to added hydrogenase, or Pt catalysts, via electron carriers such as methyl viologen and hence produce H_2. Based on this reaction, attempts have been made to produce H_2 photosynthetically by immobilized cyanobacteria. For certain species (*Chlorogloea fritschii*), pretreatment by repeated freezing and thawing in the entrapment material (polyurethane) is necessary; this does not damage the cell morphology and allows retention of H_2-evolving capacity for several days. Other algae (*Nostoc* sp. and *Mastigocladus*), grown in nitrogen-free media, can be used without any treatment with ascorbate (pH 5.5) or dithiothreitol (neutral to basic pH) as substrates by continuously removing the oxygen produced.

Whole cells of *Chlorella* sp., co-immobilized with *Clostridium butyricum* and supplemented with NADP were capable of producing H_2 for six days. For *Anabaena cylindrica*, immobilized with glass beads, hydrogen production for 30 days was described at rates 30-fold higher than determined for free-living cells, and a production of up to 13 μl H_2 mg^{-1} dry weight of *Oscillatoria* sp. immobilized in agar could be maintained over prolonged periods. As already mentioned, research into algal-cell immobilization is still in its infancy and no final conclusion can be drawn as to whether this applied algology will be of any commercial importance. It seems likely that for the immediate future, if at all, the use of immobilized algae will be restricted to the production of high-priced extracellular bioproducts. The introduction of new techniques, including permeabilization of the algal cell wall, incorporation of foreign DNA, protoplast fusion etc., may help to improve the possibilities for large-scale application of immobilized algae.

14.4.   Labeled algae

Microalgae offer a unique approach to the production of radioactively-labeled compounds because of the simple method to incorporate inexpensive labeled inorganic raw materials into complex organic molecules. The wide genetic diversity among the algal species gives rise to a plethora of compounds that can be labeled. The capability of some microalgae to grow photoautotrophically in heavy water (D_2O) has been exploited to produce various deuterated organic chemicals. Although in general many of these biochemicals may be of commercial interest, the most promising application seems to be the production of fatty acids and their derivatives. Deuterated chemicals are more stable than their hydrogenated forms: the increased mass contributed by deuterium makes the compounds more resistant to high temperatures, pressure and oxidation, properties very important for lubricants. Highly deuterated grease (99% substitution) formulations show much higher half-life times compared with the hydrogenated compounds. Thus, algae strains that can synthesize large amounts of lipids could expand the commercial scope of this technology (Chen *et al.*, 1990).

Perdeuterated fatty acids are found in low concentrations in D_2O-grown algae such as strains of *Chlorella* or *Scenedesmus*; however, this amount can be increased considerably by growing the algae under nitrogen-deficient conditions. The stress of D_2O media reduces the growth to about 80% of that in H_2O but simultaneously increases the lipid content so that the volumetric productivity is actually increased. The most abundant fatty acids are monounsaturated, which can be ozonolyzed to yield shorter saturated fatty acids that can be used for the manufacture of perdeuterated lubricants. The original saturated fatty acids can be the base for perdeuterated greases or metal soaps.

Another group of valuable labeled compounds are ^{13}C-labeled fatty acids, which can be produced by growing selected algae strains using ^{13}CO$_2$ as the sole carbon source. In contrast to D$_2$O, this has little or no isotope effect on physiological growth parameter. Conversion ratios of two parts of biomass per part of carbon consumed can be obtained. Lipids thus produced are almost completely labeled with ^{13}C. These labeled fatty acids represent a useful tool in studies on the synthesis, metabolism and bioconversion of fatty acids, especially those polyunsaturated fatty acids that are important for human nutrition. It could also be shown that considerable amounts of exogenous fatty acids in methyl-ester form can be taken up by transesterifying the fatty acids to biologically relevant lipids, to elongate and desaturate them.

15 Outlook

Analysis of the technology and economics of microalgae production demonstrates that this is a technology intermediate between agriculture, because it requires large areas to catch the necessary sunlight, and fermentation (or related microbiological processes), as it involves liquid culture of microorganisms. The advantage of algae culture over agriculture is the ability to provide much more of the desired product, because agricultural wastes such as roots, stems, stalks etc., do not accumulate. Even with the present productivities, algal cultures are more productive than conventional agricultural systems in the production of proteins or vegetable oils. Owing to the more favorable process design in algae production, this process can make better use of the annual seasons with lower light intensities. Although microalgae do not have higher photosynthetic efficiencies than conventional plants, microalgae production systems can be many times more productive in terms of output (food, fine chemicals) than agriculture. A high priority of research, by applying both genetic and physiological approaches, must be the search for genetic or adaptive characteristics that provide maximum rates of photosynthesis.

Another benefit of algal production systems is the possibility of growing algae in arid zones – unfavorable for terrestrial agriculture – as algae are able to utilize brackish or saline waters. The advantage over fermentation processes is the ability of the algae to utilize sunlight instead of organic substrates such as sugar or starch. The major disadvantages of algal cultures are their need for higher amounts of CO_2 as carbon source, problems that may be caused in culture management by a non-sterile environment, and their small size and low biomass concentration in the culture medium requiring expensive harvesting processes. The approach to use microalgae has so far been hampered by the economic constraints inherent in the required relatively sophisticated technology on one side and the production of comparatively inexpensive products such as food and feed on the other, and by critically analyzing the present situation it cannot be expected that there is much prospect for the large-scale production of algae as a source of single-cell protein in the near future. In addition, progress in algal biotechnology is often hampered by the fact that on one side algae have been cultivated without regard for any practical use and on the other isolation and application of unique algal constituents have been reported without considering the possibilities of growing and supplying these algae. It is beyond question that for the applications envisaged, additional research work is required before profitable commercial-scale production can be established. To achieve a breakthrough in microalgal technology and to promote the large-scale production of economically competitive natural compounds as well as other renewable applications from algae, it is necessary to think in terms of complementarity rather than competition at the international scientific and technological level by joint efforts of physiologists, geneticists, economists and biotechnologists. Such international concerted action has been proposed for the European Community that will cover the following main aspects.

> Evaluation of the existing know-how, research programs and industrial activities concerning microalgal technology within Europe and other parts of the world.
> Market evaluation and prospects for existing algal products and applications.
> Identification of new products and evaluation of their possible applications in consultation with relevant industries.
> Assessment of possibilities for improving algal productivity and yield of specific compounds, including screening and selection of algal strains, adjustment of growth conditions and genetic manipulation for the purpose of obtaining higher yields of biomass and special products.
> Optimization of process parameters such as illumination regime, application of stress conditions, photoheterotrophic nutrition and synchronous cultivation.
> Design, construction and optimization of large-scale cultivation equipment and advanced photo-bioreactors, both from the technical and economical point of view.
> Procedures for low-cost harvesting and processing of algal biomass and fractionation of biomass components (downstream processing).
> Assessment of the potential of microalgae for aquaculture in marine and fresh-water systems.
> Possibilities for employing algae for treatment of waste water including municipal, industrial and agricultural waste water.
> Application of microalgae for the production of biofuels, including hydrocarbons, hydrogen, combustable lipids, alcohol and methane.

It has been stated repeatedly that microalgae can produce almost anything, and under the present situation it seems that only with such an assumption is it possible to propagate algal cultivation because nobody would accept microalgae if they have no specific interest or place in contemporary economy and industry. On the other hand it has to be kept in mind that microalgae are not unique, because there is no single product that could not be obtained from other microbial or plant sources. That means that the initiative to produce microalgae must be based on economic decisions rather than on any specific necessity or personal favor for microalgae.

At present, the list of interesting and relatively well-studied algal products is short and includes pigments, polyunsaturated fatty acids, polysaccharides and phycobiliproteins. These products are produced by very few microalgal strains, i.e. *Spirulina* sp., *Dunaliella* sp. and *Porphyridium* sp. The promotion of microalgae as a health food seems to be a sort of temporary fashion and does not constitute a real product with a promising commercial future. It is a product imposed on the consumer without any evident demand or need.

One of the initial aims of algal cultivation was the production of protein-rich food supplements. Although in the course of the technological development the production cost could be reduced considerably (i.e. for *Spirulina*), the alga still cannot compete economically with conventional proteins (soya). Hence, the only possibility of promoting the commercialization of this alga is its use as a dietetic food, as a remedy and source of different chemicals such as pigments, especially phycobiliproteins, fatty acids, etc. Similar considerations apply to *Dunaliella*, which is rich in glycerol and β-carotene. Large-scale production of the two *Dunaliella* strains *D. bardawil* and *D. salina* seems to be the first successful example of the use of microalgae for the conversion of solar energy into secondary metabolites of commercial value. Whereas the glycerol content is of no immediate economic advantage, the high-priced β-carotene demonstrates the interest in algal cultivation in principle, although it has to compete with the synthetically produced pigment. The same is true for the green alga *Haematococcus*, only recently recognized as a source of the carotenoid astaxanthin, which is of special importance in aquaculture.

There is a promising future for these products because of the increasing demand for natural food-coloring agents and the strong evidence that natural β-carotene is effective as an anticancer agent.

Polysaccharides can be made from cultures of *Porphyridium cruentum*, but they have to compete with those of marine macroalgae, notably alginates, carrageenans and agarose, and with those from heterotrophic bacteria and fungi. More promising is the utilization of *Porphyridium* as a source of arachidonic acid, an essential fatty acid and precursor of prostagladins and thromboxans, because more than 30% of the fatty acids of this alga are composed of arachidonic acid. *Botryococcus*, which synthesizes unusual hydrocarbons, is another microalga with unique products and hence a candidate for large-scale production. Considering all these activities, particularly the recent establishment of several new production systems, questions of interest are the scientific and engineering aspects of such technology, the product-specific economics of algae production and the prospects of developing a profitable algae industry. There are still several specific unknowns and uncertainties and many questions to be answered concerning the achievement of low-cost production processes and development of specialty products. In conclusion, the production and commercialization of microalgae is still in its infancy. The results obtained so far in both the most important aspects, i.e. cultivation technology and production and isolation of algal constituents, have demonstrated the potential of this technology as well as the uncertainties and drawbacks. Until recently, improvement in yield and other capabilities of microalgae have been achieved by selection of algal strains and improvements in culturing and processing procedures. Very little work has been done so far by adopting a genetic-engineering approach to improve algae. The particular techniques to be used to transform specialized potentials of microalgae into realities depends on the nature of the substance of interest and the nature of the gene make-up of the organism. It is important to transform the algae genetically in a stable, facile and predictable manner, in order to explore the full biotechnological potential of these organisms. The coming years may see some interesting developments of microalgal biotechnology, with the help of genetic engineering and other new techniques that will probably give a new dimension to the utilization of microalgae.

Appendix. Addresses of algae culture collections

The following is a list of collections from where algae cultures can be obtained.

1) Sammlung von Algenkulturen (Collection of algal cultures), Pflanzenphysiologisches Institut der Universität, Nikolausberger Weg 18, 3400 Göttingen, Germany.
2) American Type Culture Collection, 12301 Parklawn Drive, Rockville, Maryland 20852, USA.
3) Culture Centre of Algae and Protozoa, Windermere Freshwater Biological Association, The Ferry House, Ambleside, Cumbria LA22 0LP, England.
4) Culture Collection of Algae, Norwegian Institute for Water Research, P.O. Box 333-Blindern, 0314 Oslo, Norway.
5) Pasteur Collection of Cyanobacteria, Unité de Physiologie Microbienne, Département de Biochimie et Génétique Moléculaire. Institut Pasteur, 28 Rue du Dr. Roux, 75724 Paris Cedex 15, France.
6) Culture Collections of Algae, Department of Botany, The University of Texas, Austin, Texas 78712, USA.

References

Al'bitskaya, O.N., Zaitseva, G.N., Rogozhin, S.V., Pakhomova, M.V., Oshanina, N.P. & Voronka, S.S. (1979). Characterization of the protein product from the *Spirulina platensis* biomass. *Prikladnaya Biokhimiya i Mikrobiologiya*, **15**, 751–4.

Anusuya Devi, M., Rajasekaran, T., Becker., E.W. & Venkataraman, L.V. (1979). Serum protein regeneration studies on rats fed on algae diets. *Nutrition Reports International*, **19**, 785–93.

Anusuya Devi, M. & Venkataraman, L.V. (1983a). Hypocholesterolic effect of the blue-green alga *Spirulina platensis* in albino rats. *Nutrition Reports International*, **28**, 519–30.

Anusuya Devi, M. & Venkataraman, L.V. (1983b). The effect of algal protein diets on the regeneration of serum and liver of protein depleted rats. *Qualitas Plantarum*, **33**, 287–97.

AOAC (Association of Official Analytical Chemists) (1990). Changes in official methods of analysis. *Journal of the Association of the Official Analytical Chemists*, **73** (1), 189.

Becker, E.W. (1983). Limitations of heavy metal removal from waste water by means of algae. *Water Research*, **17**, 459–66.

Becker, E.W. (1984). Biotechnology and exploitation of the green alga *Scenedesmus obliquus* in India. *Biomass*, **4**, 1–19.

Becker, E.W., Jakober, B., Luft, D. & Schmülling, R.M. (1986). Clinical and biochemical evaluations of the alga *Spirulina* with regard to its application in the treatment of obesity. A double-blind cross-over study. *Nutrition Reports International*, **33**, 565–74.

Becker, E.W. & Venkataraman, L.V. (1982). *Biotechnology and Exploitation of Algae – The Indian Approach*. Eschborn: German Agency for Technical Cooperation.

Becker, E.W. & Venkataraman, L.V. (1984). Production and utilization of the blue-green alga *Spirulina* in India. *Biomass*, **4**, 105–25.

Becker, E.W., Venkataraman, L.V. & Khanum, P.M. (1976). Effect of different methods of processing on the protein efficiency ratio of the green alga *Scenedesmus acutus*. *Nutrition Reports International*, **14**, 305–14.

Ben-Amotz, A. (1980). Glycerol production in the alga *Dunaliella*. In *Biochemical and Photosynthetic Aspects of Energy Production*, ed. A. San Pietro, pp. 191–208. New York: Academic Press.

Ben-Amotz, A. & Avron, M. (1980). Glycerol, β-carotene and dry algal meal production by commercial cultivation of *Dunaliella*. In *Algae Biomass*, ed. G. Shelef & C.J. Soeder, pp. 603–10. Amsterdam: Elsevier/North Holland Biomedical Press.

Ben-Amotz, A. & Avron, M. (1983). Accumulation of metabolites by halotolerant algae and its industrial potential. *Annual Review of Microbiology*, **37**, 95–119.

Ben-Amotz, A. & Avron, M. (1989). The biotechnology of mass culturing *Dunaliella* for products of commercial interest. In *Algal and Cyanobacterial Biotechnology*, ed. R.C. Cresswell, T.A.V. Rees & N. Shah, pp. 91–114. Essex: Longman Scientific & Technical.

Ben-Amotz, A., Lers, A. & Avron, M. (1988). Stereoisomers of β-carotene and phytoene in the alga *Dunaliella bardawil*. *Plant Physiology*, **86**, 1286–91.

Benemann, J.R. (1986). *Microalgae biotechnology: Products, processes and opportunities*. OMEC International Inc., Washington DC, USA.

Benemann, J.R. (1989). The future of microalgal biotechnology. In *Algal and Cyanobacterial Biotechnology*, ed. R.C. Cresswell, T.A.V. Rees & N. Shah, pp. 317–37. Essex: Longman Scientific & Technical.

Benson, A.A., Daniel, H. & Wiser, R. (1959). A sulfolipid in plants. *Proceedings of the National Academy of Sciences USA*, **45**, 1582–7.

Bizzi, A., Chiesara, E., Clementi, F., Della Torre, P., Marabini, L., Rizzi, R. & Villa, A. (1980). Trattamenti prolungati nel ratto con diete contenenti proteine di *Spirulina*, aspetti biochimici, morfologici e tossicologici. In *Prospettive della Coltura di Spirulina in Italia*, ed. R. Materassi, pp. 205–28. Firenze: Accademia dei Georgofili.

Blum, J.C. & Calet, C. (1975). Valeur alimentaire des algues spirulines pour la crooissance du poulet de chair. *Annales de la nutrition et de l'alimentaire*, **29**, 651–74.

Bock, H.D. & Wünsche, J. (1968/9). Untersuchungen über die Proteinqualität von zwei Grünalgenmehlen. *Jahrbuch für Tierernährung und Futtermittelkunde*, **6**, 544–8.

Boris, G. & Tulliez, J. (1975). Determination du 3,4-benzopyrene dans les algues spirulines produites et traitees suivant differents procedes. *Annales de la Nutrition et de l'alimentaire*, **29**, 573–9.

Borowitzka, L.J. & Borowitzka, M.A. (1989). Industrial production: methods and economics. In *Algal and Cyanobacterial Biotechnology*, ed. R.C. Cresswell, T.A.V. Rees & N. Shah, pp. 294–316. Essex: Longman Scientific & Technical.

Borowitzka, M.A. (1988). Fats, oils and hydrocarbons. In *Micro-algal Biotechnology*, ed. M.A. Borowitzka & L.J. Borowitzka, pp. 257–87. Cambridge: Cambridge University Press.

Borowitzka, M.A. & Borowitzka, L.J. (1988). *Dunaliella*. In *Micro-algal Biotechnology*, ed. M.A. Borowitzka & L.J. Borowitzka, pp. 27–58. Cambridge: Cambridge University Press.

Boudene, C., Collas, E. & Jenkins, C. (1975). Recherche et dosage de divers toxiques minéraux dans les algues spirulines de différentes origines et évaluation de la toxicité a long terme chez la rat d'un lot d'algues spirulines de provenance mexicaine. *Annales de la Nutrition et de l'Alimentaire*, **29**, 577–88.

Bourges, H., Sotomayor, A., Mendoza, E. & Chavez, A. (1971). Utilization of alga *Spirulina* as a protein source. *Nutrition Reports International*, **4**, 31–43.

Bradford, M.M. (1976). A rapid and sensitive method for the quantitation of microgram quantities of protein utilizing the principle of protein dye binding. *Analytical Chemistry*, **72**, 248–54.

Brune, H. (1982). Zur Verträglichkeit der Einzelleralgen *Spirulina maxima* und *Scenedesmus acutus* als alleinige Eiweissquelle für Broiler. *Zeitschrift für Tierphysiologie, Tierernährung und Futtermittelkunde*, **35**, 55–75.

Brune, H. & Walz, O.P. (1978). Studies on some nutritive effects of the green alga *Scenedesmus acutus* with pigs and broilers. *Archiv für Hydrobiologie, Beihefte Ergebnisse der Limnologie*, **11**, 79–88.

Burlew, J.S. (ed.) (1953). *Algal culture from laboratory to pilot plant*. Pub. no. 600, Washington D.C.: Carnegie Institution.

Butterwick, C., Heaney, S.I. & Talling, J.F. (1982). A comparison of eight methods for estimating the biomass and growth of planktonic algae. *British Phycological Journal*, 17, 69–79.

Chamorro, G. (1980). *Etude toxicologique de l'algue* Spirulina *plante pilote productrice de protéines (Spirulina de Sosa Texcoco S.A.)* UF/MEX/78/048, UNIDO/10.387.

Cheeke, P.R., Gasper, E., Boersma, L. & Oldfield, J.E. (1977). Nutritional evaluation with rats of algae (*Chlorella*) grown on swine manure. *Nutrition Reports International*, 16, 579–85.

Chen, H., Bingham, S.E., Chantler, V., Pritchard, B. & Kyle, D.J. (1990). ^{13}C-labeled fatty acids from microalgae. *Development in Industrial Microbiology*, 31, 257–64.

Chung, P., Pond, W.G., Kingsbury, J.M., Walker, E.F. & Krook, L. (1978). Production and nutritive value of *Arthrospira platensis*, a spiral blue-green alga grown on swine wastes. *Journal of Animal Sciences*, 47, 319–30.

Clement, G., Giddey, C. & Menzi, R. (1967b). Amino acid composition and nutritive value of the alga *Spirulina maxima*. *Journal of Science, Food and Agriculture*, 18, 497–501.

Clement, G., Rebeller, M. & Zarrouk, C. (1967a). Wound treating medicaments containing algae. *La France Medicale*, 29, 5279.

Clement, G. & van Landeghem, H. (1970). Spirulina, ein günstiges Objekt für die Massenkultur von Mikroalgen. *Berichte der Deutschen Botanischen Gesellschaft*, 85, 559–65.

Combs, G.F. (1952). Alga (*Chlorella*) as a source of nutrients for the chick. *Science*, 116, 453–4.

Contreras, A., Herbert, D.C., Grubbs, B.G. & Cameron, I.L. (1979). Blue-green alga, *Spirulina*, as the sole dietary protein in sexually maturing rats. *Nutrition Reports International*, 19, 749–63.

Cook, B.B. (1962). Nutritive value of waste-grown algae. *American Journal of Public Health*, 52, 243–51.

Dam, R., Lee, S., Fry, P. & Fox, H. (1965). Utilization of algae as a protein source for humans. *Journal of Nutrition*, 86, 376–82.

Davis, I.F., Sharkey, M.J. & Williams, D. (1975). Utilization of sewage algae in association with paper in diets of sheep. *Agriculture and Environment*, 2, 333–8.

Diem, K. & Letner, C. (1975). *Documenta Geigy, Wissenschsftliche Tabellen*. Stuttgart: Georg Thieme Verlag.

Dilov, C., Georgiev, D. & Bozhkova, M. (1985). Cultivation and application of microalgae in the People's Republic of Bulgaria. *Archiv für Hydrobiologie, Beihefte Ergebnisse der Limnologie*, 20, 35–8.

Dodd, J.C. (1979). Algae production and harvesting from animal waste water. *Agricultural Wastes*, 1, 23–37.

Dodd, J.C. (1986). Elements of pond design and construction. In *Handbook of Microalgal Mass Culture*, ed. A. Richmond, pp. 265–83. CRC Press Inc., Boca Raton, USA.

Dodd, J.C. & Anderson, J.L. (1977). An integrated high rate pond algae harvesting system. *Progress in Water Technology*, 9, 713–26.

Durand-Chastel, H. (1980). Production and use of *Spirulina* in Mexico. In *Algae Biomass*, ed. G. Shelef & C.J. Soeder, pp. 51–64. Amsterdam: Elsevier/North-Holland Biomedical Press.

284   *References*

El-Fouly, M.M., Mohn, F.H. & Soeder, C.J. (1985). *Joint Egyptian–German Project. Intensive protein production through microalgae.* Kernforschungsanlage Jülich GmbH, Publication 34/85.1.

Enebo, L. (1967). A methan-consuming green alga. *Acta Chemica Scandinavia*, **21**, 625–32.

Erchul, B.F.A. & Isenberg, D.L. (1968). Protein quality of various algal biomasses produced by a water reclamation plant. *Journal of Nutrition*, **95**, 374–80.

Faggi, E. (1980). Carrateristiche microbiologiche delle biomasse di *Spirulina*. In *Prospettive della coltura di* Spirulina *in Italia*, ed. R. Materassi, pp. 127–36. Firenze Accademia dei Georgifili.

FAO/WHO (1973). *Energy and protein requirement.* Report of a Joint FAO/WHO *ad hoc* Expert Committee, no. 52. Geneva: FAO.

FDA (Food and Drug Administration) Part 110 (1985). Current good manufacturing practice in manufacturing, processing, packing, or holding human food. In *Code of Federal Regulations, Title 21, Food and Drugs*, pp. 91–8. Office of the Federal Register, General Service Administration, Washington D.C.

Feldheim, W. (1972). Untersuchungen über die Verwendung von Mikroalgen in der menschlichen Ernährung. I. Ernährungsversuche mit algenhaltigen Kostformen in Thailand. *International Journal of Vitamin and Nutrition Research*, **42**, 600–6.

Fevrier, C. & Sevet, B. (1975). *Spirulina maxima* in pig feeds. *Annuales de la Nutrition et de l'Alimentaire*, **29**, 625–50.

Fisher, A.W. & Burlew, J.S. (1953). Nutritional value of microscopic algae. In *Algal Culture from Laboratory to Pilot Plant*, ed. J.W. Burlew, pp. 303–10. Washington D.C.: Carnegie Institution.

Fox, R.D. (1985). The cultivation of *Spirulina* as part of an integrated village health system. *Archiv für Hydrobiologie Beih. Ergebnisse der Limnologie*, **20**, 17–24.

Fried, A., Tietz, A., Ben-Amotz, A. & Eichenberger, W. (1982). Lipid composition of the halotolerant alga, *Dunaliella bardawil*. *Biochmica Biophysica Acta*, **713**, 419–26.

Ganowski, H., Usunova, K. & Karabashev, G. (1975). Effect of the microalga *Scenedesmus acutus* on the digestibility of the mast of calves and on some blood parameters. *Animal Science*, **12**, 74–83.

Goldman, J.C. (1979). Outdoor algal mass cultures – I. Applications. *Water Research*, **13**, 1–19.

Gornall, A.G., Bardawill, C.J. & David, M.M. (1949). Determination of serum proteins by means of the biuret reaction. *Journal of Biological Chemistry*, **177**, 751–66.

Griebsch, A. & Zöllner, N. (1970). Harnsäure-Plasmaspiegel und renale Harnsäureausscheidung bei Belastung mit Algen, einer purinreichen Eiweißquelle. *Verhandlungen der Deutschen Gesellschaft für Innere Medizin*, **77**, 173–7.

Gross, R., Gross, U., Ramirez, A., Cuadra, K., Collazos, C. & Feldheim, W. (1978). Nutritional tests with green alga *Scenedesmus* with healthy and malnourished children. *Archiv für Hydrobiologie, Ergebnisse der Limnologie, Beiheft*, **11**, 174–83.

Gross, R., Schöneberger, H., Gross, U. & Lorenzen, H. (1982). The nutritional quality of *Scenedesmus acutus* produced in a semiindustrial plant in Peru. *Berichte der Deutschen Botanischen Gesellschaft*, **95**, 323–7.

Gudin, C. & Chaumont, D. (1983). Solar biotechnology study and development of tubular solar receptors for controlled production of photosynthetic cellular biomass. In *Proc. Workshop and E.C. Contractor's Meeting in Capri.*, ed. W. Palz & D. Pirrwitz, pp. 184–93. Reidel Publishing Comp.

Gustafson, K.R., Cardellina, J.H., Fuller, R.W., Weislow, O.S., Kiser, R.F., Snader, K.M., Patterson, G.M.L. & Boyd, M.R. (1989). AIDS-Antiviral sulfolipids from cyanobacteria (blue-green algae). *Journal of the National Cancer Institute*, **81**, 1254–8.

Hedenskog, G., Enebo, L., Vendlova, J. & Prokes, B. (1969). Investigation of some methods for increasing the digestibility *in vitro* of microalgae. *Biotechnology and Bioengineering*, **11**, 37–51.

Hernandez, I.T. & Shimada, A.S. (1978). Estudios sobre el valor nutritivo del alga Espirulina (*Spirulina maxima*). *Archivos Latinoamericanos de Nutricion*, **28**, 196–207.

Heussler, P. (1985). Aspects of sloped algae pond engineering. *Archiv für Hydrobiologie, Ergebnisse der Limnologie, Beiheft*, **20**, 71–83.

Hiller, A., Plazin, J. & D.D. van Slyke (1948). A study of conditions for Kjeldahl determination of nitrogen in proteins. *Journal of Biological Chemistry*, **176**, 1401–20.

Hindak, F. & Pribil, S. (1968). Chemical composition, protein digestibility and heat of combustion of filamentous green algae. *Biologie Plantarum*, **10**, 234–44.

Hintz, H.F. & Heitmann, H. (1967). Sewage-grown algae as a protein supplement for swine. *Animal Production*, **9**, 135–41.

Hintz, H.F., Heitmann, H., Weird, W.C., Torell, D.T. & Meyer, J.H. (1966). Nutritive value of algae grown on sewage. *Journal of Animal Science*, **25**, 675–81.

Honeycutt, S.S., Wallis, D.A. & Sebba, F. (1983). A technique for harvesting unicellular algae using colloidal gas aphrons. *Biotechnology and Bioengineering Symposium*, **13**, 567–75.

Hudson, B.J.F. & Karis, I.G. (1974). The lipids of the alga *Spirulina*. *Journal of Science, Food and Agriculture*, **25**, 759–63.

Hundley, J.M. & Ing, R.B. (1956). Algae as sources of lysine and threonine in supplementing wheat and bread diets. *Science*, **124**, 536–7.

IUPAC (1974). *Proposed guidelines for testing of single cell protein destined as major protein source for animal feed.* Information Bulletin, Internation Union of Pure and Applied Chemistry, Technical Reports, no. 12. Oxford: IUPAC Secretariat.

Jassby, A. (1988). Some public health aspects of microalgal products. In *Algae and Human Affairs*, ed. C.A. Lembi and J.R. Waaland, pp. 181–202, Cambridge: Cambridge University Press.

Jaya, T.V., Scarino, M.L. & Spadoni, M.A. (1980). Charrateristiche nutrizional in vivo di *Spirulina* maxima. In *Prospettive della coltura di* Spirulina *in Italia*, ed. R. Materassi, pp. 195–208. Firenze Academia dei Georgifili.

Jüttner, F. (1982). Mass cultivation of micro-algae and phototrophic bacteria under sterile conditions. *Process Biochemistry*, **17**, 2–7.

Kashima, Y. & Tanaka, Y. (1966). *On the changes of weights and morbidity rate of common cold for the crew of the training fleet in 1966.* Report of the Central Hospital, The Self Defence Force, Japan.

Kenyon, C.N., Rippka, R. & Stanier, R.Y. (1972). Fatty acid composition and physiological properties of some filamentous blue-green algae. *Archiv für Mikrobiologie*, **83**, 216–36.

Kofranyi, E. & Jekat, F. (1967). Estimation of the biological value of food proteins, mixing of egg with rice, maize, soya, and algae. *Hoppe-Seyler's Zeitschrift für Physiologische Chemie*, **348**, 84–8.

Kondratiev, Y.I., Bychkov, V.P., Ushakov, A.S., Boiko, N.N., Klyushkina, N.S., Abaturova, E.A., Terpilovsky, A.M., Korneeva, N.A., Belyakova, M.I. & Kasatkina, A.G. (1966). The use of 50 and 100 g of dry biological bodies

containing unicellular algae in food rations of man. *Voprosy Pitanya*, **25**, 9–14.
Krishnakumari, M.K., Ramesh, H.P. & Venkataraman, L.V. (1981). Food safety evaluation: acute oral and dermal effects of algae *Scenedesmus acutus* and *Spirulina platensis* on albino rats. *Journal of Food Protection*, **44**, 934–56.
Laws, E.A., Terry, K.L., Wickman, J. & Chalup, M.S. (1983). A simple algal production system designed to utilize the flashing light effect. *Biotechnology and Bioengineering*, **25**, 2319–35.
Leveille, G.A., Sauberlich, J.W. & Shockley, J.W. (1962). Protein value and the amino acid deficiencies of various algae for growth of rats and chicks. *Journal of Nutrition*, **76**, 423–8.
Lincoln, P. & Hill, D.T. (1980). An integrated microalgae system. In *Algae Biomass*, ed. G. Shelef & C.J. Soeder, pp. 229–44. Amsterdam: Elsevier/North Holland Biomedical Press.
Lipstein, B. & Hurwitz, S. (1980). The nutritional value of algae for poultry. Dried *Chlorella* in broiler diets. *British Poultry Science*, **21**, 9–21.
Lowry, O.H., Rosebrough, N.J., Farr, A.L. & Randall, R.J. (1951). Protein measurement with the folin–phenol reagent. *Journal of Biological Chemistry*, **193**, 265–75.
Lubitz, J.A. (1962). Animal nutrition studies with *Chlorella* 71105. In *Biologistics for Space Systems Symposium*. Technical Documents Report no. AMRL-TDR-62-116, pp. 331–56.
Lubitz, J.A. (1963). The protein quality, digestibility and composition of algae, Chlorella 71105. *Journal of Food Science*, **28**, 229–41.
Mahadevaswamy, M. & Venkataraman, L.V. (1987). Bacterial contaminants in blue green alga *Spirulina* produced for use as biomass protein. *Archives of Hydrobiology*, **110**, 623–30.
Märkl, H. & Mather, M. (1985). Mixing and aeration of shallow open ponds. *Archiv für Hydrobiologie, Ergebnisse der Limnologie, Beiheft*, **20**, 85–93.
Mathews-Roth, M.M. (1985). Carotenoids and cancer prevention – experimental and epidemiological studies. *Pure and Applied Chemistry*, **57**, 717–22.
McGarry, M.G. & Tongkasame, C. (1971). Water reclamation and algae harvesting. *Journal of the Water Pollution Control Federation*, **43**, 824–35.
Merino, F., Moya, R., Rodriguez, A. & Heussler, P. (1985). Control of the parasite *Aphelidium* sp. in mass cultures of *Scenedesmus obliquus* by chemical agents. *Archiv für Hydrobiologie, Ergebnisse der Limnologie, Beiheft*, **20**, 115–23.
Mitsuda, H. (1962). *Utilisation of Chlorella for food*. Congr. Food Sci. Technol., London.
Mohn, F.-H. & Contreras, O.C. (1991). Harvesting of the alga *Dunaliella* – Some considerations concerning its cultivation and impact on the production costs of β-carotene. *Berichte des Forschungszentrum Jülich*, UNDP Project CHI/87/009.
Mokady, S. & Cogan, U. (1988). Nutritional evaluation of a protein concentrate and of carotenes derived from Dunaliella bardawil. *Journal of Science, Food and Agriculture*, **42**, 249–54.
Mokady, S., Yannai, S., Einav, P. & Berk, Z. (1980). Protein nutritive value of several microalgae species for young chickens and rats. In *Algae Biomass*, ed. G. Shelef & C.J. Soeder, pp. 655–60. Amsterdam: Elsevier/North Holland Biomedical Press.
Mokady, S., Yannai, S., Einav, P. & Berk, Z. (1979). Algae grown on waste water as a source of protein for young chickens and rats. *Nutrition Reports International*, **19**, 383–91.

Mokady, S., Yannai, S., Einav, P. & Berk, Z. (1978). Nutritional evaluation of the protein of several algae species for broilers. *Archiv für Hydrobiologie, Beihefte Ergebnisse der Limnologie*, **11**, 89–97.

Müller-Wecker, H. & Kofranyi, E. (1973). Einzeller als zusätzliche Nahrungsquelle. *Hoppe-Seyler's Zeitschrift für Physiologische Chemie*, **354**, 1034–42.

Nakashima, M.J. (1989). Extraction of light filth from *Spirulina* powders and tablets: colloraborative study. *Journal of the Association of the Official Analytical Chemists*, **72**, 451–3.

Nakaya, N., Homma, Y. & Goto, Y. (1988). Cholesterol lowering effect of *Spirulina*. *Nutrition Reports International*, **37**, 1329–37.

Narasimha, D.L.R., Venkataraman, G.S. Duggal, S.K. & Eggum, O. (1982). Nutritional quality of the blue green alga *Spirulina platensis Geitler*. *Journal of Science, Food and Agriculture*, **33**, 456–60.

Necas, J. & Lhotsky, O. (1967). *Annual Report Algology Laboratory, Trebon*. Czechoslovakian Academy of Science Institute of Microbiology, Trebon.

Nigam, B.P., Venkataraman, L.V., Hoffman, N.Q. & Grunewald, T. (1990). Color reduction in foods containing microalgae. *Journal of Food Science Technology*, **3**, 136–9.

Nigam, B.P., Venkataraman, L.V., Seibel, L.V., Brummer, W., Seiler, J.M. & Stephan, H. (1985). Addition of algae to wheat bread and wheat extrudates. *Getreide, Mehl und Brot*, **39**, 53–6.

Oswald, W.J. (1988). Large-scale algal culture systems (engineering aspects). In: *Micro-algal biotechnology*, ed. M.A. Borowitzka & L.J. Borowitzka, pp. 305–94. Cambridge: Cambridge University Press.

Pabst, W. (1974). Die Proteinqualität einiger Mikroalgenarten, ermittelt im Rattenbilanzversuch: *Scenedesmus, Coelastrum, Uronema*. *Zeitschrift für Ernährungswissenschaft*, **13**, 73–80.

Pabst, W., Payer, H.D., Rolle, I. & Soeder, C.J. (1978). Multigeneration feeding studies in mice for safety evaluation of the microalga *Scenedesmus acutus*. *Food Cosmetics and Toxicology*, **16**, 259–64.

Payer, H.D. (1977). *Annual Report of the Thai–German Algal Project*, Kasetsart University, Bangkok, Thailand.

Payer, H.D., Pithakpol, B., Nguitragool, M., Prabharaksa, C., Thananunkul, D. & Chavana, S. (1978). Major results of the Thai–German Microalgae Project at Bangkok. *Archiv für Hydrobiologie, Ergebnisse der Limnologie*, **11**, 41–55.

Payer, H.D. & Runkel, K.H. (1978). Environmental pollutants in fresh water algae from open-air mass cultures. *Archiv für Hydrobiologie, Beihefte Ergebnisse der Limnologie*, **11**, 184–9.

Piorreck, M., Baasch, K.-H. & Pohl, P. (1984). Biomass production, total protein, chlorophylls, lipids and fatty acids of freshwater green and blue-green algae under different nitrogen regimes. *Phytochemistry*, **23**, 207–16.

Pirt, S.J. (1983). Maximum photosynthetic efficiency: a problem to be resolved. *Biotechnology and Bioengineering*, **25**, 1915–22.

Pirt, S.J., Lee, Y.-K., Richmond, A. & Watts-Pirt, M. (1980). The photosynthetic efficiency of *Chlorella* biomass growth with reference to solar energy utilization. *Journal of Chemical Technology and Biotechnology*, **30**, 25–34.

Pohl, P., Kohlhase, M., Krautwurst, S. & Baasch, K.H. (1987). An inexpensive inorganic medium for the mass cultivation of freshwater microalgae. *Phytochemistry*, **26**, 1657–9.

Pohl, M., Kohlhase, M. & Martin, M. (1986). Pilot scale axenic mass cultivation of

microalgae. I. Development of the biotechnology. *Planta Medica*, **52**, 416–17.
Powell, R.C., Nevels, E.M. & McDowell, M.E. (1961). Algae feedings in humans. *Journal of Nutrition*, **75**, 7–12.
Priestly, G. (1976). Algal proteins. In *Food from Waste*, ed. G.G. Birch, H.J. Parker & J.T. Worgan, pp. 114–38. London: Applied Science Publishers.
Protein Advisory Group (PAG) (1972). *Guidelines for pre-clinical testing of novel sources of proteins. No. 6*. Protein Advisory Group of the United Nations.
Raven, J.A. (1988). Limits to growth. In *Micro-algal Biotechnology*, ed. M.A. Borowitzka & L.J. Borowitzka, pp. 331–56. Cambridge: Cambridge University Press.
Reddy, V.R., Reddy, C.V., Varadarajulu, P. & Reddy, G.V. (1978). Utilization of algae protein (*Scenedesmus acutus*) in chicken rations. *Indian Poultry Gazette*, **62**, 67–70.
Richmond, A. & Becker, E.W. (1986). Technological aspects of mass cultivation – a general outline. In *Handbook of Microalgal Mass Culture*, ed. A. Richmond, pp. 245–63. CRC Press, Boca Raton.
Robinson, P.K., Mak, A.L. & Trevan, M.D. (1986). Immobilized algae: a review. *Process Biochemistry*, **21**, 122–7.
Rolle, I. & Pabst, W. (1980). Über die cholesterinsenkende Wirkung der einzelligen Grünalge *Scenedesmus acutus* 276–3a. *Nutrition and Metabolism*, **24**, 291–301.
Ross, E. & Dominy, W. (1989). The nutritional value of dehydrated, blue-green algae (*Spirulina platensis*) for poultry. *Poultry Science*, **69**, 794–800.
Rydlo, O. (1973). Verwendung einiger Mikroalgen in der praktischen Pharmazie. *Pharmazie*, **6**, 146–7.
Safar, F. (1975). Algotherapie. *Prakt. Lek (Praha)*, **55**, 641–8.
Saito, T. & Oka, N. (1966). Clinical applications of *Chlorella* preparations. *Shinryo Shinyaku*, **3**, 3–9.
Setlik, I., Veladimir, S. & Malek, I. (1970). Dual purpose open circulation units for large scale culture of algae in temperate zones. I. Basic design considerations and scheme for pilot plant. *Algological Studies (Trebon)*, **1**, 111–64.
Shi, D.J. & Hall, D.O. (1988). The *Azolla–Anabaena* association: historical perspective, symbiosis and energy metabolism. *Botanical Review*, **54**, 353–86.
Shinohara, K., Okura, Y., Koyano, T., Murakami, H., Kim, E. & Omura, H. (1986). Growth-promoting effects of an extract of a thermophilic blue-green alga, *Synechococcus elongatus* var. on human cell lines. *Agricultural and Biological Chemistry*, **50**, 2225–30.
Simmer, J. (1969). Outdoor mass cultivation of *Scenedesmus quadricauda (Turp.) Breb.* in South Bohemia. In *Studies in Phycology*, ed. B. Fott, pp. 293–304. Stuttgart: E. Schweitzerbartsche Verlagsbuchhandlung.
Smidsrød, O. & Skjåk-Bræk, G. (1990). Alginate as immobilization matrix for cells. *Trends in Biotechnology*, **8**, 71–8.
Soeder, C.J. (1979). Verschiedene Single Cell Protein Typen und ihre biochemischen Eigenschaften. *Fette-Seifen-Anstrichmittel*, **81**, 352–7.
Spoehr, H.A. & Milner, H.W. (1949). The chemical composition of *Chlorella*. Effect of environmental conditions. *Plant Physiology*, **24**, 120–49.
Stengel, E. & Soeder, C.J. (1975). Control of photosynthetic production in aquatic ecosystems. In *Photosynthesis and Productivity in Different Environments*, ed. J.P. Cooper, pp. 645–60. Cambridge: Cambridge University Press.
Tamiya, H. (1957). Mass culture of algae. *Annual Review of Plant Physiology*, **8**, 309–34.

Tanaka, K., Konishi, F., Himeno, K., Taniguchi, K. & Nomoto, K. (1984). Augmentation of antitumor resistance by a strain of unicellular green algae, *Chlorella vulgaris. Cancer Immunology and Immunotherapy*, **17**, 90–4.

Thananunkul, D., Reungmanipaitoon, S., Prasomsup, U., Pongjiwanich, S. & Klafs, H.J. (1977). The protein quality of algae produced in Thailand as determined by biological assays with rats. *Food (IFRPD), Bankok*, **10**, 200–9.

Torzillo, G., Pushparaj, B., Bocci, F., Balloni, W., Materassi, R. & Florenzano, G. (1986). Production of *Spirulina* biomass in closed photobioreactors. *Biomass*, **11**, 61–74.

Tsukada, O., Kawahara, T., & Miyachi, S. (1977). Mass culture of *Chlorella* in Asian countries. *Biological Solar Energy Conversion*, ed. A. Mistui, pp. 363–6. New York: Academic Press.

Vasquez, V. & Heussler, P. (1985). Carbon dioxide balance in open air mass culture of algae. *Archiv für Hydrobiologie, Ergebnisse der Limnologie, Beiheft*, **20**, 95–113.

Venkataraman, L.V. & Becker, E.W. (1985). *Biotechnology and Utilization of Algae. The Indian Experience*. Department of Science and Technology, New Delhi, India.

Venkataraman, L.V., Becker, E.W. & Khanum, P.M. (1977b). Supplementary value of the proteins of alga *Scenedesmus acutus* to rice, ragi, wheat and peanut proteins. *Nutrition Reports International*, **15**, 145–55.

Venkataraman, L.V., Becker, E.W., Khanum, P.M. & Mathew, K.R. (1977a). Short-term feeding of alga *Scenedesmus acutus* processed by different methods: growth pattern and histopathological studies. *Nutrition Reports International*, **15**, 231–40.

Venkataraman, L.V., Becker, E.W., Rajasekaran, T. & Mathew, K.R. (1980). Investigations on toxicology and safety of algal diets in albino rats. *Food and Cosmetics Toxicology*, **18**, 271–5.

Vonshak, A. (1987). Strain selection of *Spirulina* suitable for mass production. *Hydrobiologia*, **151/152**, 75–7.

Vonshak, A. (1986). Laboratory techniques for the cultivation of microalgae. In *Handbook of Microalgal Mass Culture*, ed. A. Richmond, pp. 117–45. Boca Raton: CRC Press.

Waslien, C.J., Calloway, D.H., Margen, S. & Casta, F. (1970). Uric acid levels in men fed with algae and yeast as protein sources. *Journal of Food Science*, **35**, 294–8.

World Health Organization (WHO) (1972). *Evaluation of certain food additives and the contaminants mercury, lead, and cadmium*. World Health Organization, Technical Report Series 505.

Yamaguchi, M., Hwang, H.G., Kawaguchi, K. & Kandatsu, M. (1973). Metabolism of ^{15}N-labelled *Chlorella* protein in growing rats with special regard to the nutritive value. *British Journal of Nutrition*, **30**, 411–24.

Yannai, S. & Mokady, S. (1985). Short-term and multi-generation toxicity tests of algae grown in wastewater as a source of protein for several animal species. *Archiv für Hydrobiologie, Ergebnisse der Limnologie, Beiheft*, **20**, 173–80.

Yannai, S., Mokady, S., Sachs, K., Kantorowitz, B. & Berk, Z. (1980). Certain contaminants in algae and in animals fed algae-containing diets, and secondary toxicity of algae. In *Algae Biomass*, ed. G. Shelef & C.J. Soeder, pp. 757–66. Amsterdam: Elsevier/North Holland Biomedical Press.

Yannai, S., Mokady, S., Sachs, K., Kantorovitz, B. & Berk, Z. (1979). Secondary toxicology and contaminants of algae grown on waste water. *Nutrition Reports International*, **19**, 391–402.

Yap, T.N., Wu, J.F., Pond, W.G. & Krook, L. (1982). Feasibility of feeding *Spirulina maxima*, *Arthrospira platensis* or *Chlorella* sp. to pigs weaned to a dry diet at 4 to 8 days of age. *Nutrition Reports International*, **25**, 543–52.

Yoshida, M. & Hoshii, H. (1982). Nutritive value of new type of *Chlorella* for poultry feed. *Japanese Poultry Science*, **19**, 56–8.

Yoshino, Y., Hirai, Y., Takahashi, H., Tamamoto, N. & Yamazaki, N. (1980). The chronic intoxication test on *Spirulina* product fed to Wistar rats. *Japanese Journal of Nutrition*, **38**, 221–6.

Yu, M.K., Hsia, I.T., Li, S.H., Liu, K.S. & Wang, C.W. (1959). The artificial precipitation of unicellular green algae in mass cultures. *Acta Hydrobiologica Sinica*, p. 488.

Zahradnik, J. (1968). Outdoor cultivation. *Annual Report of the Laboratory of Experimental Algology and Department of Applied Algology for the Year 1967*, pp. 141–6.

Zarrouk, C. (1966). *Contribution à l'étude d'une cyanophycée. Influence de divers facteurs physiques et chimiques sur la croissance et la photosynthèse de Spirulina maxima*. Ph. D. Thesis, University of Paris.

Index